COMPUTER ASSISTED STUDIES OF CHEMICAL STRUCTURE AND BIOLOGICAL FUNCTION

Computer Assisted Studies of Chemical Structure and Biological Function

ANDREW J. STUPER

Rohm and Haas Company
Spring House, Pennsylvania

WILLIAM E. BRÜGGER

International Flavors and Fragrances, Inc.
Union Beach, New Jersey

PETER C. JURS

Pennsylvania State University
University Park, Pennsylvania

A Wiley-Interscience Publication
JOHN WILEY & SONS
New York · Chichester · Brisbane · Toronto

Copyright © 1979 by John Wiley & Sons, Inc.

All rights reserved. Published simultaneously in Canada.

Reproduction or translation of any part of this work beyond that permitted by Sections 107 or 108 of the 1976 United States Copyright Act without the permission of the copyright owner is unlawful. Requests for permission or further information should be addressed to the Permissions Department, John Wiley & Sons, Inc.

Library of Congress Cataloging in Publication Data

Stuper, Andrew J., 1950-
   Computer assisted studies of chemical structure and biological function.

   "A Wiley-Interscience publication."
   Includes bibliographical references and index.
   1. Structure-activity relationship (Pharmacology) 2. Pattern recognition systems.
I. Brügger, William E., joint author. II. Jurs, Peter C., joint author. III. Title.

QP906.S75S77   612'.015   78-12337

ISBN 0-471-03896-2

Printed in the United States of America

10 9 8 7 6 5 4 3 2 1

Preface

Studies at the interfaces between biology, medicine, and chemistry are generating new information at an ever increasing rate and are changing man's perspective on life. New disciplines have grown out of mergers among more classical disciplines—biochemistry, biophysics, microbiology, medicinal chemistry, and the pharmaceutical sciences. Medicinal research has become an inately polydisciplinary field of study. While the study of biology at the organ level and the tissue level has shed a great deal of light on how chemicals interact with biological systems, fundamental progress demands that further understanding be achieved at the molecular level. Ultimately, drugs and other bioactive materials interact with biological systems at the molecular level through chemical and/or physical interactions that must obey well-known laws, such as the laws of thermodynamics and microreversibility. However, the exceeding complexity of these interactions and the biological systems involved—even when the organism is as "simple" as a bacterium—belies efforts to deal with the fundamental processes at the molecular level. Thus, for the most part, we are ignorant about the detailed mechanisms of action of bioactive compounds.

Since the unraveling of the fundamental interactions between bioactive compounds and the biological systems at the molecular level seems to be a long term goal, and since there are short term pressures to solve problems such as the conquest of a specific disease, development of symptom alleviating drugs, and the development of safe and effective herbicides and pesticides, accommodations must be reached in the interests of expediency. A number of methods have been developed to investigate the relationships between the molecular structures of bioactive compounds and their biological activity profiles. These methods typically do not provide direct evidence regarding the mechanism of interaction responsible for the activity, but they do provide methods for dealing with important, immediate, practical problems. The field of drug design or drug development has grown with the development of these techniques.

The past decade has witnessed an explosive growth in interest in rational methods for the investigation of structure–activity relations. A number of methods have been developed and presented. Hansch analysis, the method for using extrathermodynamic linear free energy relationships, grew out of the application of physical organic chemistry techniques and multivariate statistical analysis methods to problems from medicinal chemistry. Quantum mechanical methods were applied to structure–activity problems as soon as the computational techniques would allow. In other fields computer organization and manipulation of chemical and biological data grew and gained sophistication and utility. A field that has come to be known as chemical structure information handling was developed and generated a literature of its own. Computer hardware and software became accessible and economically feasible for scientists to use routinely. Sophisticated statistical computer packages became widely and easily available to the practicing scientist, and nonstatistical and nonparametric data analysis methods came to be used as well.

In the early 1970s a number of these scattered disciplines began to converge as it was recognized that the capabilities of the techniques were complementary. Thus new methods for attacking old problems were developed. This volume has been written to describe the results of bringing together techniques drawn from several fields to generate new methods for studying chemical structure–biological activity relations.

To perform investigations in the structure–activity area requires the handling and manipulation of reasonably large quantities of data. Certain of the necessary operations require nearly continual guidance on the part of the scientist to ensure that the experimental operations are performed correctly. These considerations, along with economic ones, focus attention on the use of laboratory computers that can be run directly by the scientist. Additionally, the operations necessary to do structure–activity studies dictate the overall format of the computer software to be used. Such a software system must be modular, since any given section of the system is large and thus the overall system would be too large for any feasible computer. The system must be interactive, which dictates the necessity for graphics hardware and support software. It must support and use bulk storage capabilities, that is, magnetic tapes or discs. It must be flexible for ease of expansion, contraction, or revision. Finally, it must be easy to use and capable of dealing with realistically large problems.

In this volume we describe an approach to the structure–activity problem and a computer system developed to implement this approach. Methods drawn from the field of chemical structure information handling have been combined with molecular modeling and pattern recognition techniques to develop an integrated systems approach to the structure–activity problem.

The volume begins by introducing the methodologies that have been developed for studying structure–activity relations: the Hansch linear free energy related method using physicochemical parameters; the Free-Wilson method; and quantum mechanical methods. The principles of pattern recognition are introduced and the applicability of these methods for a variety of chemical and biological research areas is shown. The methods of chemical structure information handling, chemical structure storage, and the use of these methods for the computer generation of molecular structure descriptors are described. An introduction to molecular mechanics and three-dimensional model building and the use of these methods for the generation of geometrical molecular structure descriptors are presented. The following chapter describes linear discriminant functions and how they can be used in conjunction with the previously described computer generated descriptors to find invariances in the structures of molecules of similar biological activity. Also described is a method for the investigation of the relative importance of molecular structure features with respect to the classification being attempted. The ADAPT (*a*utomatic *d*ata *a*nalysis using *p*attern recognition *t*echniques) computer system that has been implemented to incorporate these techniques is then discussed. The final chapter of the volume describes the application of these techniques to studies of central nervous system agents in two specific studies dealing with sedatives and tranquilizers and with a set of barbiturates. Also, it describes the application of these techniques to studies of chemical communicants, including studies of musk odorant compounds, trigeminal nerve stimulants, and compounds of other odor quality classes.

ANDREW J. STUPER
WILLIAM E. BRÜGGER
PETER C. JURS

Spring House, Pennsylvania
Union Beach, New Jersey
University Park, Pennsylvania
August 1978

Contents

Chapter 1
INTRODUCTION   1

   STUDIES OF STRUCTURE–ACTIVITY RELATIONS   1
   HANSCH ANALYSIS: LINEAR FREE ENERGY
   RELATIONS   2
   FREE-WILSON ADDITIVITY MODEL   15
   QUANTUM MECHANICAL METHODS   17
   APPLICATIONS OF PATTERN RECOGNITION   19
   REFERENCES   24

Chapter 2
PATTERN RECOGNITION PRINCIPLES   30

   BASIC PATTERN RECOGNITION METHODS   31
   PREPROCESSING   38
   CLASSIFICATION   51
   CLUSTERING CONCEPTS   56
   REFERENCES   58

Chapter 3
CHEMICAL STRUCTURE INFORMATION HANDLING:
MOLECULAR DESCRIPTOR DEVELOPMENT   62

   BASIC CONCEPTS OF MOLECULAR STRUCTURE
   CODING   62
   WISWESSER LINE NOTATION   64
   CONNECTION TABLES   66
   STRUCTURE ENCODING   68
   MOLECULAR STRUCTURE DESCRIPTORS:
   TOPOLOGICAL   73
   MOLECULAR MODELING AND GEOMETRIC
   DESCRIPTORS   83

SUMMARY	92
REFERENCES	93

Chapter 4

PATTERN RECOGNITION: LINEAR DISCRIMINANT FUNCTIONS · 95

THE LINEAR LEARNING MACHINE	96
GRADIENT DESCENT AND THE LEAST SQUARES ALGORITHM	100
NEAREST NEIGHBOR CLASSIFICATION	104
LIMITATION OF NONPARAMETRIC LINEAR CLASSIFIERS	105
THE VARIANCE METHOD OF FEATURE SELECTION	112
REFERENCES	125

Chapter 5

A SOFTWARE SYSTEM TO IMPLEMENT COMPUTER ASSISTED STRUCTURE–ACTIVITY STUDIES: THE ADAPT SYSTEM · 126

THE STRUCTURE FILE MANAGEMENT SYSTEM	129
CLASS DEVELOPMENT	131
DESCRIPTOR GENERATION	131
ACTIVE DATA SET: FORMATION, FEATURE SELECTION, AND CLASSIFICATION	135
SUMMARY	136

Chapter 6

DRUG STRUCTURE–ACTIVITY RELATION STUDIES · 138

APPLICATION TO PSYCHOTROPIC AGENTS	139
APPLICATION TO BARBITURATES	151
REFERENCES	168

Chapter 7

STRUCTURE-ACTIVITY STUDIES OF OLFACTORY STIMULANTS · 170

BASIC ANATOMY AND PHYSIOLOGY OF THE NOSE	171
THEORIES OF OLFACTION	173

Contents

ANALYSIS OF MUSK ODORANTS	176
ANALYSIS OF TRIGEMINALLY ACTIVE COMPOUNDS	192
REFERENCES	212

Appendix
A LIST OF THE MUSK ODORANTS — **215**

INDEX — **217**

CHAPTER 1

Introduction

STUDIES OF STRUCTURE-ACTIVITY RELATIONS

The rational search for compounds with a desirable biological activity profile requires knowledge of the relationship between molecular structure and biological activity. Structure activity correlations are important in the development of pharmacological agents, as well as in the development of other biologically active compounds such as herbicides, pesticides, olfactory stimulants, and gustatory stimulants. Knowledge of these relations can also lead to an understanding of the toxic, mutagenic, or carcinogenic effects of a wide number of compounds.

The discovery and design of biologically active compounds is a field that has undergone widespread and well-documented (1–6) changes in the past decade with the introduction of new techniques and perspectives. The discovery process (3,6–8) can be decomposed into two general approaches: (1) the attempt to find new "lead" compounds and (2) the attempt to fully exploit existing lead compounds. A lead compound is a molecule that has the biological activity of interest, although the activity may be weak. The search for new lead compounds can proceed in any one or combination of the following ways:

1. Isolation, purification, and identification of compounds from natural products, including plant sources, animal sources, and microorganism sources. Examples of drugs found in this way include antibiotics, alkaloids, steroids, and cardiac glycosides.
2. Following up leads generated by therapeutic folklore or folk medicine.
3. Testing of metabolites or molecular modifications of metabolites of known drug compounds.
4. Fundamental studies of biochemical systems.
5. The investigation of side effects of experimental or clinically used drugs.
6. Mass screening of chemical compounds for possible biological activity.
7. Organic synthesis aimed at the production of bioactive compounds.

The procedures for rationally exploiting a lead are much more fully developed than those for development of new leads as is reflected by the voluminous literature pertaining to the former [one bibliography (9) contains 392 references and covers only the literature through November 1974]. Studies aimed at the further exploration of lead compounds consist of manipulating the molecular structures of the compounds to (*a*) bring about changes in their biological activity profile, such as eliminating bad side effects, enhancing their potency, and combining effects, to obtain selectivity with respect to an organism or species, and (*b*) develop substitutes for known biologically active compounds.

The basic assumption in the exploitation of a lead compound is that similar compounds have similar action. Thus small perturbations in structural composition should result in small perturbations in biological activity. Through systematic alteration of the structure of an active molecule, a more desirable molecule can be found. Unfortunately, the number of possible alterations for even a small compound is often astronomical. Therefore, the medicinal chemist is faced with the prospect of judging which few of a large number of possible compounds should be synthesized. A further complication lies in the fact that structural similarity is not the primary factor governing drug action. The biological effectiveness of a molecule is governed by a combination of that compound's electronic nature, steric nature, and transport properties. Structural alterations have various effects on each of these factors and, thus similarity with respect to structure is not readily apparent. Even so, it remains that alterations in the factors that govern activity are available only through the modification of structure. The choice of which modifications to make is assisted by models that reliably indicate the direction in which synthetic efforts may be most fruitfully pursued. Several methods have been developed that attempt to provide this assistance. These methods fall into the following major categories: (1) use of the semiempirical linear free energy (LFER) or extrathermodynamic model proposed by Hansch and co-workers; (2) use of the additivity or Free-Wilson model; and (3) use of quantum mechanically based approaches. We provide a short discussion of each of these methods in the sections that follow.

HANSCH ANALYSIS: LINEAR FREE ENERGY RELATIONS

Hansch analysis is a powerful technique for use in optimizing the activity of a lead compound. There exist many review articles addressing all facets of the application of this technique (1–23). The approach is based on the formation of an empirical model of drug action that uses linear free energy-related parameters as the independent variables. Thus the method is often referred to as a linear free energy related approach. The basic assumption in this method is that all the factors involved in variations in biological activity arising from the modifications of molecular structures within a congeneric series can be correlated with concomitant changes in physicochemical parameters. Furthermore, all physicochemical factors that relate to the transport and receptor interaction can be broken down into hydrophobic, electronic, and steric components. The contribution of each of these components is expressed with substituent constants that represent the difference in properties between the parent compound of the series and the compound being coded.

Hydrophobic properties are modeled using the logarithm of the partition coefficient, P, between a lipid model system, normally n-octanol and water.

The partition coefficient is used to develop an index, π, that expresses the difference between that compound and the parent of the series. The form of this π equation is

$$\pi = \log P_s - \log P_0 \tag{1.1}$$

where P_s is the partition coefficient for the substituted compound and P_0 is the partition coefficient for the parent compound in the congeneric series.

Electronic properties are accounted for by Hammett constants or other electronic substituent constants. Steric properties are coded using steric substituent parameters such as the Taft steric constant, E_s.

A large number of studies have been conducted using a linear model of the form:

$$\log \frac{1}{C} = k_1 \pi + \rho \sigma + k_2 \tag{1.2}$$

Here π and σ are determined for several compounds in the series and the constants k_1 and k_2 are found by regression analysis. C represents the molar concentration of the member of a congeneric series necessary to produce a defined biological response. Examples of the biological responses used include LD_{50}, that dose that kills 50% of the animals tested; ED_{50}, the dose of antagonist necessary to reduce by 50% the response to a standard dose of agonist; MIC, minimum growth inhibitory concentration; and I_{50}, the molar concentration of an inhibitor that reduces the rate of an enzyme catalyzed reaction by one-half.

In 1964 Hansch and Fujita (24) combined two hypotheses with the Hammett equation (25) to arrive at the most widely used equation in structure–activity studies. They postulated that the rate of biological reaction (BR) is equal to the product of three terms: A, the probability of the drug molecule reaching the site of action in a given time interval; C, the extracellular molar concentration of the drug; and k_x, the reaction rate of the receptor–drug interaction:

$$\text{rate of BR} = \frac{d(\text{response})}{dt} = ACk_x \tag{1.3}$$

The product of A and C was called the effective concentration of the drug and is the drug concentration accumulating at the critical site.

The second hypothesis consisted of assuming the A term to be Gaussian:

$$A = f(\pi) = a \exp\left[\frac{-(\pi - \pi_0)^2}{b}\right] \tag{1.4}$$

where a and b are constants. This equation can be substituted into the previous one to get

$$\frac{d(\text{response})}{dt} = Ck_x \, a \exp\left[\frac{-(\pi - \pi_0)^2}{b}\right] \tag{1.5}$$

Because drugs are commonly tested by varying the concentration of the drug until a particular response is obtained, the term $d(\text{response})/dt$ can be replaced with a constant. The exponential equation can be simplified by taking the logarithm and collecting constants to obtain

$$\log\frac{1}{C} = -k\pi^2 + k'\pi\pi_0 - k''\pi_0^2 + \log k_x + k''' \tag{1.6}$$

Use of the Hammett equation and collection of constants produces the equation

$$\log\frac{1}{C} = -k\pi^2 + k'\pi\pi_0 - k''\pi_0^2 + \rho\sigma + k'''' \tag{1.7}$$

Since the term π_0 refers to the parent compound in the congeneric series, it is constant, so the equation can be altered to give

$$\log\frac{1}{C} = -k\pi^2 + k'\pi + \rho\sigma + k'' \tag{1.8}$$

The constants of this equation are determined by regression analysis on a set of known data.

The above equations represent an attempt to justify the use of a certain linear model. However, the formation of a usable model is the goal and many other models are found in practice. The purpose of the present approach is to derive parameters that relate to the hydrophobic, steric, and electronic changes caused by making modifications to the structure of the molecule.

These parameters are used to form a first or second order linear model that accounts for the observed changes in the activity. This is an empirical approach whose justification lies in the fact that the predictive equations developed are quite reliable. In the next few pages, we present an overview of the types of parameters most often used to develop these models, the methods used to develop the models, and the applications of the technique.

Hydrophobic Parameters

The most frequently used hydrophobic parameters for structure–activity studies have been the partition coefficient P and the related parameter π, which is defined in equation 1.1. The underlying assumption of the use of log P or π is that a nonpolar solvent can be used as a reference system for interactions of drug molecules with lipid biophases.

The hydrophobic parameters log P and π are usually measured using n-octanol and water as the solvent system. A number of reasons have been presented in the literature for the choice of n-octanol, but the major reason now is that there exists a large volume of data measured using this solvent. A table containing partition coefficient values for several thousand compounds has been published (26) and the number now measured is over 10,000 (22). Since a 1975 publication has reviewed the reasons for the selection of n-octanol as a reference system (27), further discussion is not given here.

For many systems approximate log P or π values can be calculated by assuming that contributions for substituents are additive for the substructural components of the molecule. An approach to the calculation of lipophilicity directly from molecular structures has been described by Nys and Rekker (28-30). The method depends on identifying a set of hydrophobic fragmental constants (f values) that represent the contribution of individual substructures to the overall lipophilicity of a molecule. The value of log P is then defined as

$$\log P = \sum a_j f_j \qquad (1.9)$$

where f_j is the hydrophobic fragmental constant and a_j is the number of times the fragment occurs in the structure. The f_j values were found by using multiple regression analysis on a large set of molecules with known lipophilicities. The lipophilicity of a new molecule can then be computed by summing the f values for the fragments found in the molecule. Another additive method has been reported by Leo et al. (31). The methods of Nys and Rekker and of Leo et al. have been shown to be based on the same assumptions (32). Prediction of partition coefficients directly from molecular conformation produced by a sophisticated molecular mechanics routine has been reported (33). It has been found that π or log P values are not additive in molecules in which intramolecular interactions are not negligible. These include heterocyclic molecules, sterically crowded molecules, and conformationally flexible molecules. The most reliable value for the partition coefficient is obtained by measuring it experimentally.

The inclusion of a π^2 term in the linear free energy equations was originally an empirical postulate by Hansch and Fujita (24). Subsequently, two different arguments have been advanced supporting the use of the quadratic relationship.

Penniston et al. (34) presented an argument based on solving the differential equations expressing the kinetics of a model containing alternating aqueous and lipid phases.

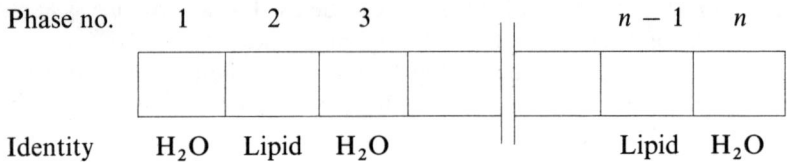

The rate at which drug molecules pass from any water phase to the adjacent lipid phase is k; the reverse rate is ℓ. The partition coefficient is k/ℓ. The concentration of drug in phase i is A_i. Then

$$\frac{dA_1}{dt} = (\ell A_2 - k A_j) \tag{1.10}$$

$$\frac{dA_{2i}}{dt} = -2\ell A_{2i} + k(A_{2i-1} + A_{2i+1}) \tag{1.11}$$

$$\frac{dA_{2i+1}}{dt} = -2k A_{2i+1} + \ell(A_{2i} + A_{2i+2}) \tag{1.12}$$

$$\frac{dA_{n-1}}{dt} = -(\ell + m)A_{n-1} + k A_{n-2} \text{ for } n \text{ odd} \tag{1.13}$$

$$= -(k + m)A_{n-1} + \ell A_{n-2} \text{ for } n \text{ even} \tag{1.14}$$

$$\frac{dA_n}{dt} = m A_{n-1} \tag{1.15}$$

m is the reaction rate of the receptor–drug interaction. This set of differential equations was solved by numerical integration. Since A_1/A_n does not depend on the absolute value of A_1^0, arbitrary values were used for A_1^0. Values of k and ℓ were chosen so that the product of k and ℓ was unity.

The equations were solved numerically for a variety of values of P at a fixed time (10 units) after dosage. A plot was made of log C (the logarithm of the concentration in the last compartment) versus log P, and an approximately quadratic shape was observed. Quadratic functions of the form

$$\log \frac{1}{C} = k_1 \log^2 P + k_2 \log P + k_3 \qquad (1.16)$$

gave good fits to the sets of points. In view of the simplicity of the model the agreement between the points and the quadratic curves was considered to be good.

Dearden and Townend (35) have investigated this partitioning model more extensively. They have shown that behavior other than the parabolic relationship can be observed, and they have cited examples in the drug structure–activity literature that conform to their theoretical predictions.

McFarland (36) presented an alternative derivation also using a model employing alternating lipid and aqueous phases. It is based on a simple probability argument. He found that the probability of a drug reaching a receptor site is a function of the partition coefficient (k/ℓ) and the number of intervening interfaces between the first aqueous phase and the receptor site. The form of the equation derived was

$$\text{probability} = \frac{(k/\ell)^{n/2}}{(k/\ell + 1)^n} \qquad (1.17)$$

For values of n greater than unity the maximum value of this function occurs for $k/\ell = 1.00$. The shape of the function for other values of k/ℓ is close to quadratic.

The inclusion of a $(\log P)^2$ term in a structure–activity relation implies the existence of an optimum degree of hydrophobic character ($\log P_0$) for the set of compounds studied. The value can be found from $[\partial \log(1/C)]/[\partial \log P]$. Discussions have appeared in the structure–activity literature regarding the importance of $\log P_0$ values, especially for CNS active agents (22).

In addition to the $\log P$ or π parameters, hydrophobicity can be represented by other physicochemical constants. An example is the chromatographic parameter R_M defined as $\log[(1/R_f) - 1]$, where R_f is the ratio of the distance moved by the compound to the distance moved by the solvent in a chromatography experiment (e.g., see references 37 and 38). Retention times derived from liquid chromatography experiments can be related to the log P values for the n-octanol/water system (e.g., see references 39 and 40). Tables of hydrophobicity parameters have been presented in the literature (5,14).

Electronic Parameters

The Hammett free energy related parameter σ was the first choice for an electronic parameter. This parameter was borrowed from physical organic chemistry, where it was developed for generating correlations between chemical reactivity and structure (24). The reaction used in the original derivation was the ionization of substituted benzoic acids. The basic equation is

$$\log \frac{k_s}{k_0} = \rho\sigma \qquad (1.18)$$

where σ is the substituent constant. It is a measure of the electron-donating or electron-withdrawing power of the substituent. The reaction parameter, ρ, is a measure of the sensitivity of the rate constant to changes in σ. It is specific for a given reaction and depends on the effect of reaction conditions such as temperature, reagent, and solvent. ρ is defined to be unity for the standard reaction, ionization of benzoic acids in water at the standard temperature of 25°C. The Hammett equation uses two sets of σ's for meta and para substitution. The reference substituent is hydrogen, so the values for σ_{meta} and σ_{para} are defined to be zero.

Substituents with positive σ values are electron withdrawing and those with negative σ values are electron donors. Reactions characterized by positive ρ values are aided by electron-withdrawing substituents on the benzene ring, and reactions with negative ρ values are aided by electron-donating substituents.

A large number of modified Hammett parameters have been developed for application to different compound classes. These include σ_o, σ_m, and σ_p, which are for ortho, meta, and para substitution, respectively; σ_p^- for substituent groups that are electron acceptors by resonance; and σ^* for aliphatic compounds (with σ^* for —CH_3 defined to be zero). The electronic substituent parameter σ can be factored into a field term and a resonance term called F and R by Swain and Lupton (41). A compilation comparing σ with F and R values recalculated by a corrected procedure has been published (42).

Many other electronic parameters have been employed in linear free energy relation studies. A table in a review by Purcell et al. (5) gives a total of over 30 experimental parameters and over 20 theoretical quantum mechanical parameters that have been reported. A list of parameters is also included in the review by Verloop (12).

Steric Parameters

Both intramolecular and intermolecular steric interactions must be considered. The most widely used steric parameter is the E_s constant defined by Taft (43) for application to correlations of structures of aliphatic esters with rates of hydrolysis reactions. Modifications of the Taft steric parameter have been developed, including $E_s^{o,m}$ and E_s^p for ortho/meta or para substitution, respectively, and E_s^c, a corrected steric parameter. A review of the E_s substituent constant and a compilation of values has been published (44). Other representations of steric effects used include molar volume, van der Waals radius, and interatomic distance. A table of some steric parameters reported for use in structure–activity relation studies is included in a review by Purcell et al. (5). Verloop et al. have shown that steric constants calculated from van der Waals radii were effective as structure–activity parameters (45). Simon has developed an alternative approach to the study of the steric fit of molecules using a minimal steric difference concept (46).

Miscellaneous Parameters

A large number of other physicochemical parameters have been employed in linear free energy related studies. Many of these parameters provide information about the molecular structure of the compound directly. For example, the molecular weight and number of certain atom types have been used. Molar refractivity, thought to be a measure of polarizability, is a parameter that has been used in several studies (22). Purcell et al. give a table of several miscellaneous parameters that have been reported (5). A recent review by Hansch (22) discusses a number of miscellaneous parameters, spectroscopic constants, and indicator variables. Indicator variables are specifically included to code the presence of a substructural unit in the molecule (21). Also, studies have appeared wherein parameters drawn from experiments were mixed with substructural or indicator variables (e.g., see ref. 47).

Regression Analysis and Statistical Parameters (48–50)

Once a set of independent variables has been chosen, the next step in the analysis is to perform a least squares regression analysis. Commonly reported statistical parameters include the standard deviation of the fit, the multiple regression coefficient, the F statistic, and the explained variance.

The biological data are assumed to be more variable and less accurately determined than the physicochemical parameters. Therefore, the biological data are assigned the role of the dependent variable and the physicochemical parameters are considered as independent variables in the regression. After a fit has been generated, the resulting statistical parameters are used to judge which equations best explain the data set.

For the sake of simplicity and generality we call the independent variables in the regression equation $x_j, j = 1, 2, \ldots, m$, and formulate the regression equation as

$$y_i^* = f(x_j) \qquad j = 1, 2, \ldots, m \qquad (1.19)$$

or

$$y_i^* = a_0 + a_1 x_1 + a_2 x_2 + \cdots + a_m x_m + \varepsilon \qquad (1.20)$$

where y_i^* is the calculated value of the dependent variable for the ith compound in the set and ε is an error term. The values of x_j for the different compounds are assumed to be normally distributed with a mean of μ and a variance of σ^2. The set of independent variables, x_j, comprises the physicochemical parameters for the ith compound. Since the value of the intercept, a_0, is given by

$$a_0 = \bar{y} - a_1 \bar{x}_1 - a_2 \bar{x}_2 - \cdots - a_m \bar{x}_m \qquad (1.21)$$

the regression equation can be rewritten as

$$y_i^* = \bar{y} + a_1 x_1 + a_2 x_2 + \cdots + a_m x_m \qquad (1.22)$$

where the new x_j values equal the original ones minus the mean, for example, $x_1(\text{new}) = x_1(\text{old}) - \bar{x}_1$.

The values of the parameters a_j, the partial regression coefficients, are found by the standard least squares procedure of minimizing the sums of squares of residuals:

$$\sum_{i=1}^{n} (y_i^* - y_i)^2 \qquad (1.23)$$

with respect to the a_j's, where y_i is the experimentally observed value of the dependent variable for the ith compound. The coefficient a_j is the amount by which y changes on the average when x_i changes by one unit and all the other x_j's remain constant.

Given a set of data and the model equation, the a_j values are found as shown below.

We form a matrix of coefficients calculated from the data being fit:

$$\mathbf{D} = \mathbf{X}'\mathbf{X} = \begin{bmatrix} \sum x_1^2 & \sum x_1 x_2 & \cdots & \sum x_1 x_m \\ \sum x_2 x_1 & \sum x_2^2 & \cdots & \sum x_2 x_m \\ \vdots & \vdots & & \vdots \\ \sum x_m x_1 & \sum x_m x_2 & \cdots & \sum x_m^2 \end{bmatrix} \qquad (1.24)$$

The vector $\mathbf{A}$ forms the desired regression coefficients:

$$\mathbf{A} = \begin{bmatrix} a_1 \\ a_2 \\ \vdots \\ a_m \end{bmatrix} \qquad (1.25)$$

and the vector $\mathbf{Y}$ holds the dependent variables:

$$\mathbf{Y} = \begin{bmatrix} y_1 \\ y_2 \\ \vdots \\ y_n \end{bmatrix} \qquad (1.26)$$

Hansch Analysis: Linear Free Energy Relations

Then, in matrix notation,

$$\mathbf{XA} = \mathbf{Y} \tag{1.27}$$

or by substituting equation 1.24

$$\mathbf{DA} = \mathbf{X'Y} \tag{1.28}$$

$\mathbf{C}$, the inverse of the matrix $\mathbf{D}$, can be obtained numerically:

$$\mathbf{C} = \{c_{ij}\} = (\mathbf{X'X})^{-1} \tag{1.29}$$

Then

$$\mathbf{A} = \mathbf{CX'Y} = (\mathbf{X'X})^{-1}\mathbf{X'Y} \tag{1.30}$$

and by simple matrix multiplication the values for the regression coefficients a_j are obtained. The reason for solving explicitly for the inverse matrix, $\mathbf{C}$, is that the individual entries in this matrix, c_{ij}, are used to find the standard errors of the regression parameters.

Once a set of a_j values has been found, the quality of the fit of the calculated y_i^* values to the y_i values can be expressed in a variety of ways.

Typically, the figures n, s, r and r^2 (51) are reported along with the actual equation.

n is the number of data points, or compounds, used in deriving the regression equation parameters. s is the standard deviation or standard error of the regression:

$$s^2 = \frac{\sum_{i=1}^{n}(y_i^* - y_i)^2}{(n-k)} \tag{1.31}$$

where k is the number of fit parameters, $k = m + 1$. Here m is the number of independent variables, and k is one greater than m because of the constant term in the regression equation. The quantity s is a measure of how closely the regression fits the data.

r is the multiple correlation coefficient:

$$r^2 = \frac{\sum_{i=1}^{n}(y_i^* - \bar{y}_i)^2}{\sum_{i=1}^{n}(y_i - \bar{y}_i)^2} \tag{1.32}$$

where

$$\bar{y}_i = \frac{1}{n}\sum_{i=1}^{n} y_i \tag{1.33}$$

r^2 is that fraction of the sum of the squares of the deviations of the set of y_i values from their mean that is accounted for by the regression. The maximum value is unity, and the minimum value is zero. As more parameters are included in the equations, that is, as m is increased, r^2 increases. One deficiency of the correlation coefficient is that its value depends on the absolute values of the regression coefficients, the a_j's (52).

To test whether all the regression parameters are different from zero with some statistical significance, the F-test is used. The F statistic is defined as

$$F = \frac{\sum (y_i^* - \bar{y}_i)^2/(k - 1)}{\sum (y_i - \bar{y}_i)^2/(n - k)} \tag{1.34}$$

or

$$F = \frac{(n - k - 1)r^2}{k(1 - r^2)} \tag{1.35}$$

where k is the number of regression parameters, $k = m + 1$. The F statistic computes the ratio of the correlation coefficient for two different degrees of freedom. After F is calculated a statistical table is used to determine the probability that attaches to the values of n, k, and F. When the F value calculated from the regression equation is greater than the critical value of F (for a given value of probability) indicated in the table, then the correlation is significant at that probability level. The derivation of the F statistic assumes that the dependent variable, here $\log(1/C)$, is normally distributed and random.

Another statistical parameter that has been reported is the explained variance, v.

$$v = 1 - \frac{\sum (y_i^* - \bar{y}_i)^2/(n - k)}{\sum (y_i - \bar{y}_i)^2/(n - 1)} \tag{1.36}$$

v is a measure of the fraction of the total variance of the set of y_i values that is accounted for or "explained" by the regression equation. It is dependent on the degrees of freedom in the regression equation. If v is small, for example, less than 0.5, then most of the variation in the set of y_i values must be due to variables not included in the regression.

The standard deviations, or standard errors, for the regression parameters are found from the set of equations

$$s_{a_j} = s\sqrt{c_{jj}} \tag{1.37}$$

where c_{jj} is the jth entry on the diagonal of the inverse matrix formed during calculation of the a_j values. The probability of each regression parameter being nonzero can be tested using the standard t-test and forming the ratios

$$t_j = \frac{a_j}{s_{a_j}} \tag{1.38}$$

and looking in t-distribution tables under $(n - k)$ degrees of freedom to find the associated probability of the parameter a_j being significant.

Confidence limits can be calculated for each parameter in the regression equation using the Student t-test. The confidence limit for a parameter a_j is calculated from the equation

$$\text{CL} = a_j \pm ts_{a_j} \qquad (1.39)$$

where s is the standard error of the parameter and t is a function of the desired probability and also of n. The t value is obtained from statistical tables.

It is common to use procedures for regression where the independent variables are successively added to a regression equation while the concomitant changes in the statistical criteria are monitored (forward selection). The aim is to include only the smallest number of variables necessary to provide a statistically significant correlation. Automated computer programs are widely available to perform this operation (e.g., see reference 53). In the forward selection method the best single variable is selected from those available. Then a computation is done to find which of the remaining variables provides the largest additional improvement in the statistical parameters being monitored. The variable identified as best is selected as the second variable. Then the remaining variables are checked to see which will provide the largest additional improvement, and so on. This continues until no variable remains that will provide sufficient additional improvement to warrant inclusion. In addition to selecting variables one at a time, some procedures involve checking at each step to see if any of the previously included variables can be discarded. These variable selection procedures are based on the assumption that the best variables identified one at a time will also form the best set of variables. This is not necessarily true, especially for variables with substantial intercorrelation.

Newer methods for performing regression analyses utilizing, for example, Furnival and Wilson's (54) method of "leaps and bounds" regression or ridge regression (55) should aid in the development of meaningful regression equations. Furnival and Wilson describe methods that can be used to investigate all possible regressions using all subsets of a supplied set of independent variables. This approach has been used by workers in drug design to ensure that no interesting grouping of variables is overlooked in a Hansch analysis. The methods available for selection of variables for linear regression have recently been reviewed by Hocking (56). Lewi (57) has described a number of more advanced methods of multivariate statistics, focusing on the use of principal components analysis.

A number of pitfalls arise in using multiple regression techniques, and discussions of these have appeared in the structure–activity literature (12,58,59). A set of five criteria that can be used to judge which is the best equation for a problem has been presented (60). The data set being used must contain a sufficient number of compounds to avoid correlations due only to chance and to support statistical significance. A rule of thumb often applied is that five or six compounds per degree of freedom are needed in the regression equation to assure significance. Each physicochemical parameter included in the regression equation must have a sufficient spread of values within the data set. Care must be taken not to include a compound in the data set that differs greatly from the other compounds with respect to one parameter but not the other parameters. Such a compound would have undue influence on the final regression equation. Each variable included in the model should be tested for significance, that is, validated. Care must be taken if physicochemical parameters are used for studies in systems that are different from the ones in which they were determined. The presence of intercorrelations among the independent variables must be noted; the use of such parameters is acceptable only with certain precautions. The qualitative model developed must be consistent with what is known about the physical-organic and medicinal chemistry of the process under investigation.

Applications

The published application studies using Hansch analysis are too numerous to list or even survey completely here. Rather, we have chosen to refer to a number of review and summary articles that in turn reference the primary publications. Tute (12) has given examples of Hansch analysis, including penicillins, sulfanilamides, adrenergic blocking agents, and studies of sweet tasting compounds. Verloop (14) has presented tables giving references to individual articles, with one table for each of the following areas: permeability and transport, hydrophobic binding, enzymatic conversions, inhibition of enzymes, paramacological activity, chemotherapeutical activity, herbicidal and plant-growth-regulating activity, pesticidal activity, and miscellaneous. Cammarata and Rogers (13) have presented a wide variety of specific relations developed using cellular systems, simple intact organisms, and animals. Dunn (15) and Cramer (21) have reported a number of examples drawn from drug design and pharmacology. Hansch (22) has presented a large number of examples, including those from the areas of simple proteins, cell membranes, nerve potential, interactions of ligands and purified enzymes, organelles, viruses and microorganisms, CNS and local anesthetic agents, acute toxicity, steroids and adrenergic agents, pesticides, cancer antitumor drugs, and environmental studies.

FREE-WILSON ADDITIVITY MODEL

In the Free-Wilson or additivity model the biological response (BR) observed for a compound is defined to be equal to the sum of the overall activity of the substituent groups plus the overall average activity:

$$A_i = \mu + \sum_j a_{j,p} \tag{1.40}$$

$a_{j,p}$ is the contribution to the activity from the jth substituent in the pth position on the parent structure, and A_i is the standard biological response for compound i in the series. This equation involves the assumption that every time a particular substituent group appears at the same place in the molecule it makes a constant contribution toward determining the overall activity of the compound. The values of the individual substituent group contributions are calculated using multiple linear regression analysis. To perform the regression, only the biological activities and the molecular structures are needed; no physicochemical parameters are employed.

Three slightly different, but related, formulations of the Free-Wilson method have been presented in the literature. The original model presented by Free and Wilson (61) is given by

$$\text{BR} = \mu + \sum_{ij} G_{ij} X_{ij} \tag{1.41}$$

A modified version was developed by Cammarata (62):

$$\text{BR} = \mu_\text{H} + \sum_{ij} a_{ij} X_{ij} \tag{1.42}$$

and another modified version was presented by Fujita and Ban (63):

$$\text{BR} = \mu_0 + \sum_{ij} a_{ij} X_{ij} \tag{1.43}$$

In these equations μ is the overall average of biological activity values for the series; G_{ij} is the activity contribution of the substituent X_i in position j (i.e., $X_{ij} = 1$ if the substituent X_i is in position j; otherwise, $X_{ij} = 0$); a_{ij} is the group contribution of the substituent X_i in position j, based on the definition that a_H equals zero; μ_H is the observed biological activity of the unsubstituted compound (all substituents are H); and μ_0 is the theoretically predicted biological activity of the unsubstituted compound (all substituents are H). A detailed comparison of these three models has been presented by Kubinyi and Kehrhahn (64).

The Free-Wilson model and the Hansch linear free energy relation equation containing only first order terms,

$$\log \frac{1}{C} = k_1 \pi + \rho\sigma + k_2 E_s + k_3 \qquad (1.44)$$

have been shown to be theoretically equivalent (62). Example data sets (65) have been used to study this relationship as well.

The Hansch approach and the Free-Wilson approaches to structure–activity relation studies can be combined to form mixed or hybrid relations such as

$$\log \frac{1}{C} = \sum a_j + k_1 \pi + \rho\sigma + \mu \qquad (1.45)$$

where some of the substituents are coded using a_j values and some are coded using physicochemical parameters. This mixed approach is mentioned above in connection with the addition of "indicator variables" to Hansch regression equations. A detailed discussion of the utility of this mixed approach has been presented by Kubinyi (66), and comparisons of the two methods have been made by Craig (67).

To perform a Free-Wilson analysis on a set of compounds, a series of equations is set up and the parameters $a_{j,p}$ are calculated using least squares procedures. The same statistical coefficients described above with reference to Hansch analysis are used here. If the statistical parameters obtained are satisfactory, and the additivity assumption is validated, then the constants can be used to regenerate the biological activities of those compounds comprising the data set. Outliers can quickly be identified. More importantly, the constants can be used to predict the activities of compounds with all combinations and permutations of the given substituents. The relative effects on biological activity due to the various substituent groups in the various positions can be ranked by investigating the sizes of the constants $a_{j,p}$.

The most important disadvantage of Free-Wilson analysis is the large number of variables needed to describe all substituents. In addition, problems must be designed to avoid matrices with singularities. Thus use of the Free-Wilson approach presents a dilemma to the medicinal chemist: either a large number of derivatives must be made, or the number of substituents and/or placements must be limited. The choice is obviously determined by the factors governing the specific study in progress.

Compared to the Hansch analysis or quantum mechanical approaches to structure–activity studies, there are relatively few literature reports of Free-Wilson analyses. Cammarata and Rogers (13) have reported several applications. Dunn (15) has described work on antimalarials and antihypertensives. Purcell et al. (5) have given a number of examples drawn from the primary literature. In several recent papers on the Free-Wilson approach, Kubinyi (64,65) has presented a number of pharmaceutical examples and compared Hansch and Free-Wilson analyses of several data sets. Craig lists some applications (67).

QUANTUM MECHANICAL METHODS

Structure–activity relations are being studied by quantum mechanical methods, largely molecular orbital theory. Several recent reviews have summarized the work being done in this area (68–75). Quantum mechanical techniques are used to pursue two goals—calculation of theoretical parameters that can be correlated with activity and the determination of preferred conformations of bioactive molecules. A discussion of the quantum mechanical methods applicable to structure–activity problems has been presented by Kaufman and Koski (71) and by Richards and Black (72).

Methods that have been applied to biological structure–activity studies include the Hückel (HMO) method, which considers only π electrons, that is, delocalized electrons, and which can be used for studies of conjugated coplanar molecular systems; the extended Hückel (EHT) method, which takes into account all the valence electrons in a molecule and allows calculation of the total energies of the different conformations and thus energies of barriers to internal rotations; the iterative extended Hückel method (IEHT); complete neglect of differential overlap (CNDO) and its modifications such as CNDO/2; intermediate neglect of differential overlap (INDO); modified INDO (MINDO) and its updated versions MINDO/2 and MINDO/3; perturbative configuration interactions using localized orbitals (PCILO), *ab initio* LCAO-MO-SCF calculations, linear combinations of atomic orbitals to form molecular orbitals by a self-consistent field technique. An excellent discussion of the semiempirical methods is given in reference 76.

Quantum mechanical calculations provide computed indices that reflect the electronic structure of molecules. The electronic charge distribution at each atom and between the atoms is generated. Relative values of physicochemical parameters that can be calculated include resonance energies, dipole moments, ionization potentials, electron affinities, charges on atoms, transition energies, molecular conformations, the energy of the highest occupied molecular orbital (HOMO), the energy of the lowest empty molecular orbital (LEMO), superdelocalizability, and frontier electron density. The energy of the HOMO correlates with ionization potential and thus is related to the ability of a molecule to donate an electron. The energy of the LEMO correlates with electron affinities and is related to the ability of the molecule to accept an electron. Superdelocalizability indicates the stabilization energy that can be gained in the formation of a complex with another molecule. For an electron-donating reaction it is large when the electron density in the HOMO is large. A table in reference 5 gives more than 20 theoretically derived indices that have been utilized as electronic parameters in multiple linear regression models.

One of the major pieces of information obtained from quantum mechanical calculations is what possible low energy conformations of a molecule can exist. In this approach the total energy of the molecule is calculated as a function of bond rotation (with bond lengths and angles held constant). However, different preferred conformations can be predicted by different calculations depending on the values used for molecular parameters such as bond lengths and bond angles and also on which quantum mechanical method was used. There is evidently no agreement on which of the empirical or semiempirical methods is most accurate. The safest way to use MO methods is to compare conformations for closely related series of structures using the same method of calculation.

One use for these conformational calculations is as follows. One particular MO method can be used to calculate the preferred conformation of each of several potent drugs of differing structure but common pharmacological activity. The drug receptor is mapped by considering those portions of the various drug molecules that present nearly identical patterns of like-charged atoms. Using this method, Kier (77,78) mapped receptors for acetylcholine, nicotine, serotonin, histamine, steroids, and α-adrenergic agents.

The calculations of quantum mechanics are done on lone gas phase molecules without consideration of solvent effects. Solvation may be an important contributor to conformational stability for some bioactive molecules. Intramolecular interactions that could be present in the gas phase, for example, hydrogen bonding, may not be present in the solution phase where solvent molecules can provide hydrogen bonding.

Quantum chemical approaches have been taken to the study of a variety of bioactive compounds. Kier (68) has described research in several areas, including pharmacology (antimalarials, anesthetics, tranquilizers, analgesics, antihypertensives, and anticonvulsants), herbicides and pesticides, carcinogens, and hallucinogens, and he has described research dealing with the mapping of pharmacophores. Dunn's review (15) mentions several pharmacological examples. Neely (69) has described studies of acetylcholine, glucopyranose derivatives, nucleosides, amino acids and polypeptides, hallucinogens, antihypertensives, and herbicides. Green et al. (70) have discussed quantum chemical studies of cholinergic compounds, adrenergic compounds, dopamines, serotonins, histamines, hallucinogens, neuroleptics, and other CNS active compounds. Richards and Black (72) have mentioned several single compound conformation studies in their review article. Christoffersen (73) has reviewed over 300 papers dealing with the application of quantum mechanical methods to approximately 25 pharmacological classes of compounds under subheadings of agents affecting nerve function, cardiovascular function, endocrine function, anti-infective agents, and miscellaneous. In another paper Christoffersen and Angeli (75) have reviewed work that appeared during 1975 dealing with quantum mechanical studies of adrenergics, analgesics, antibiotics, anticancer drugs, antiallergy drugs, cholinergics, enzyme–substrate interactions, and so forth.

APPLICATIONS OF PATTERN RECOGNITION

The methods of pattern recognition, described fully in the following chapters of this volume, have been applied to a number of studies in the fields of drug design, agricultural chemicals, and chemical communicants. Since a relatively small number of papers have appeared describing this work, these reports are discussed in the following few pages. Applications of pattern recognition to drug design have also been reviewed by Kirschner and Kowalski (79).

A study by Ting et al. has appeared in which an attempt was made to correlate the mass spectra of a set of 30 sedatives and 36 tranquilizers with their biological activities (80). The mass spectral peak intensities in 30 mass to charge positions were used to represent each compound. Several pattern recognition data analysis methods were used. The compounds could be classified into these two categories with a high degree of accuracy. Two papers (81,82) subsequently appeared that questioned the independence of the sets of compounds used and the total number of compounds used, and thus questioned the significance of the results obtained. Of the 66 compounds used, over half of the sedatives were barbiturates and over half of the tranquilizers were phenothiazines.

In another attempt to show correlations between the mass spectra of bioactive compounds and their activity classifications, a set of 16 analgesics and 16 antispasmodics were used (83). Each compound was represented by the mass spectral peak intensities in 262 mass to charge positions. This large number of peaks was reduced by using principal components analysis and nonlinear mapping. The data set was analyzed using distance classification methods and a linear learning machine method. Classification success rates between 70 and 100% were reported for different combinations of feature reduction and data analysis. The data set used would seem to be open to the question of the degree of variety of structures in the data set.

A study by Hiller et al. (84) used perceptrons, a forerunner of linear learning machines, to analyze a set of 48 alkyl- and alkoxyalkyl-substituted 1,3-dioxanes. Each compound was coded by its substitution pattern using five primitive substructural entities: H, O, CH, CH_2, and CH_3. A binary classifier was developed to divide the data set into the two categories of active or inactive as an antagonist to corasol. Recognition percentages of 85 to 90 for the training sets and approximately 70 for the prediction sets were obtained.

Martin et al. (85) reported the results of a study of compounds that inhibited monoamine oxidase. Twenty compounds were represented by lipophilicity and steric parameters. Discriminant analysis was used to separate the data set into four groups—inactive, slightly active, moderately potent, most potent—and into two groups—inactive or slightly active as opposed to moderately potent or most potent. The results of classification were discussed.

A second study utilizing the same data set as the mass spectral correlation work has appeared (86). The paper reported the results of applying linear learning machines, Fisher discriminants, and a number of cluster analysis techniques to the binary problem of separating sedatives from tranquilizers. The compounds were represented by a set of 46 basic substructural fragments called augmented atoms. The number of features per compound was reduced to 16 by using Fisher ratios and probabilities. Classification success rates between 84 and 94 % of the compounds were reported.

Kowalski and Bender (87) reported a structure–activity study of antitumor drugs. Two hundred drugs tested by the National Cancer Institute for activity in the solid tumor Adenocarcinoma 755 screening system were represented by 20 structural features each. The data set was analyzed using three pattern recognition methods with classification results of approximately 90 % correct responses. The compounds were classified with respect to a threshold of activity to separate semiquantative data into two categories. Two subsequent papers have appeared (88,89) criticizing this work by questioning the selection of the data set and of the set of features used to represent each compound.

Chu et al. (90) have reported the results of a study dealing with antitumor drugs. A set of 138 structurally diverse compounds were each represented by three types of descriptors; atom-centered fragments, "heteropath" fragments, and ring nuclei. Each compound had been tested in an ependymoblastoma study on mice. The two classes of compounds used were those that gave a 25 % increase in life span of the mice and those compounds that failed to do so. From a total of 421 unique features per compound, a subset of 51 was chosen, and this set was analyzed with the nearest neighbor method and with a linear learning machine. Success rates of 83 and 93 % were reported. A set of 24 unknowns was then predicted and reported.

Stuper and Jurs (91) used linear learning machines to study a set of 140 tranquilizers and 79 sedatives. Each set of compounds contained a wide variety of structural types. A total of 69 descriptors was generated for each compound, and the data were separable into the two classes based on this set of descriptors. Feature selection reduced the number of descriptors per compound to approximately 40 with no loss in separability. Predictive abilities on unknowns of similar structural types of approximately 90% were reported. This study is described in greater detail in Chapter 6.

Cammarata and Menon (92) applied several pattern recognition methods to a set of compounds of accepted therapeutic utility. The compounds were coded by matching them with appropriate templates, using numerical codes to obtain a set of features expressing the type of substructure present in certain portions of each molecule. A correlation matrix of features was submitted to factor analysis, and the largest eigenvectors were plotted for visual analysis. This analysis was performed for a set of 13 pressor agents and for a set of 43 compounds classified as antihistamines, anticholinergics, analgesics, antidepressants, antipsychotics, and anti-Parkinsonian agents. For the second data set the molar refractivity was also used as a descriptor. The eigenvector plots were presented and discussed.

Menon and Cammarata (93) have reported the results of another study similar to the first one, but using a set of 39 compounds classified as alpha- and beta-adrenergic agents, cholinergic agents, and central nervous system stimulants. Principal components analysis and display were used. For a subset of the compounds a second coding, transformation, and display were necessary. It was concluded that the displays of the three principal components allowed the major pharmacological groups to be identified.

Stuper et al. (94,95) have reported the development of a computer software system, called ADAPT, designed for use in SAR studies using pattern recognition. The system has been used to study a set of barbiturates in an effort to classify them with respect to a set of thresholds. A brief description of the study is given. The same study of barbiturates is presented in greater detail in a review by Stuper and Jurs (96). A data set of 160 5,5'-disubstituted barbiturates with a wide variety of cyclic substituents was coded using 46 numerical descriptors, including fragments, substructures, environmental descriptors, and the molecular connectivity index. Linear discriminant functions were developed that could dichotomize the data set with respect to several thresholds separating the longer- from the shorter-acting compounds. Feature selection was used to focus on the subset of most important molecular features. Predictive abilities of approximately 94% were reported. This study is described in greater detail in Chapter 6.

The papers described to this point focus on the pattern recognition analysis of sets of compounds for classification or prediction. There have also been papers that focus on the development of descriptors from the molecular structures. Several of these are discussed in the following paragraphs.

Cramer et al. (97) have presented work on the coding of compounds using substructures for "substructural analysis." Seven hundred compounds of known biological activity in an antiarthritic–immunoregulatory test were studied. A series of statistical quantities was calculated for each substructural fragment for the set of 770 compounds. A series of 77 studies was performed wherein 10 compounds were considered to be unknowns and were predicted by comparison with the remaining 760 compounds using several probabilistic measures.

Adamson and Bush have reported studies (98–100) of coding of compounds using substructures consisting of atom centered fragments, that is, a central atom, the bonds it forms, and the nonhydrogen atoms to which it is bonded. In the first work (98) 79 penicillins were studied by multiple linear regression to sets of substructural descriptors. In the second study (99) 39 local anesthetics were classified by calculating similarity or dissimilarity coefficients between pairs of structural diagrams and then applying cluster analysis to the results. In the third study (100) the 39 local anesthetics were examined using fragments derived from their connection tables, which were analyzed by regression analysis. Quantitative predictions of the minimum blocking concentration of the 39 compounds were in good agreement with the observed values.

Brugger et al. (101) reported the development of a set of computer routines for the generation of molecular structure descriptors from the connection tables of compounds. Descriptors of two basic types—topological and geometrical—were discussed. Topological descriptors included fragments, substructural descriptors (which code the presence or absence of particular, explicitly defined, substructures), and environmental descriptors, which code the immediate surroundings of an atom center of interest. The geometrical descriptors were derived from a strain-energy minimized three-dimensional structure, and they coded the size and shape of the molecule.

Dierdorf and Kowalski (102) described a method for the calculation of descriptors from three-dimensional models of compounds. A molecule to be described was oriented using principal component rotation. Then a matrix of atomic information for the molecule was submitted to principal components analysis, and the eigenvectors of the matrix were used for pattern recognition analysis. Forty six derivatives of *para*-dimethylaminoazobenzene and 39 benz(*a*)anthracene derivatives were classified with linear discriminant functions as carcinogenic or noncarcinogenic with 80 and 92% success rates, respectively. A major limitation of this molecular representation is that physical significance is lost in passing through the transformations.

Soltzberg and Wilkins (103,104) have reported studies on the representation of molecular structures using methodology borrowed from electron diffraction and termed molecular transform. The molecular transform is computed directly from the three-dimensional atomic coordinates of the molecule being represented. A set of 114 tranquilizers and 72 sedatives was studied using linear discriminant functions. Predictive abilities of approximately 90% were reported. The possibility of extracting geometric prototypes for each activity class was discussed.

Gund (105) has reported the development of methodology that allows the searching of three-dimensional molecular structures for pharmacophores. A table of proposed pharmacophoric patterns taken from a literature search was described, and a selection of examples including an antileukemic pattern, an analgesic pattern, and a prokariotic ribosomal transpeptidase inhibitor pattern were described.

In addition to studies of structure–activity relations of pharmaceuticals, the methods of pattern recognition have been applied in a preliminary way to studies of chemical communication.

McGill and Kowalski (106) have reported a study on the number of descriptors necessary to account for odor quality. Forty-seven compounds were studied; each compound was represented by 43 descriptors, including UV, IR, NMR data, results from CNDO calculations, and so forth.

Brugger and Jurs (107) reported a study on musk compounds. A data set of 60 musk odorants and 240 nonmusk odorants was coded with computer generated structural descriptors and then analyzed using a linear learning machine. Thirteen descriptors were found that were sufficient to classify all 300 compounds into the two classes; these were then used to predict the classes of a set of unknowns with a high degree of accuracy.

Hansch et al. (108) have reported the use of cluster analysis in an application to drug design. Hierarchical clustering was used to study relationships among a set of 90 substituents, each of which was represented by a selection of physicochemical parameters. Several groupings of the available physicochemical parameters were used, and the results of each study were discussed in detail. The objective was to provide a method for aiding the drug designer in choosing a subset of all available substituents that would still give a good, representative selection of all possible structures.

White and Lewinson (109) have reported the use of a newly developed clustering method for analysis of data sets including the relation between chemical structures and biological responses. The clustering procedure is based on a probabilistic measure of similarity generated entirely from the given data set. Its application to a set of 38 compounds forming an analogue series was presented.

Hodes et al. (110) have reported the use of a statistical heuristic method for the screening of very large files of compounds for effective antitumor drugs. The method is based on principles similar to those used previously by Cramer et al. (97). A set of new compounds is ranked according to predicted activity by comparing their group average properties to these same properties in the main file of compounds.

REFERENCES

1. E. J. Ariens, *Drug Design*, Vols. 1–7, Academic, New York 1971–1976.
2. A. Burger, *Medicinal Chemistry*, Part I, Wiley-Interscience, New York 1970.
3. B. Bloom and G. E. Ullyot (Eds.), *Drug Discovery*, Advances in Chemistry Series, No. 108, American Chemical Society, Washington, D.C., 1971.
4. W. V. Valkenburg (Ed.), *Biological Correlations—The Hansch Approach*, Advances in Chemistry Series, No. 114, American Chemical Society, Washington, D.C. 1972.
5. W. P. Purcell, G. E. Bass, and J. M. Clayton, *Strategy of Drug Design*, Wiley–Interscience, New York, 1973.
6. Y. C. Martin, *Quantitative Drug Design. A Critical Introduction*, Dekker, New York, 1978.
7. E. J. Ariens, A General Introduction to the Field of Drug Design, in *Drug Design*, Vol. I., E. J. Aneus (Ed.), Academic, New York, 1971.
8. A. Goldstein, L. Aranow, and S. M. Kalman, *Principles of Drug Action: The Basis of Pharmacology*, 2nd ed., Wiley, New York, 1974, pp. 741ff.
9. Science Information Services Department, Franklin Institute Research Laboratories, *Structure–Activity Correlation Bibliography: With Subject and Author Index*, PB-240 658/5 GA, March 1975.
10. C. Hansch, A Quantitative Approach to Biochemical Structure–Activity Relationships, *Acc. Chem. Res.*, **2**, 232 (1969).
11. C. Hansch, Quantitative Structure–Activity Relationships in Drug Design, in *Drug Design*, Vol. I, E. J. Ariens (Ed.), Academic, New York, 1971.
12. M. S. Tute, Principles and Practice of Hansch Analysis: A guide to Structure–Activity Correlation for the Medicinal Chemist, in *Advances in Drug Research*, Vol. 6, N. J. Harper and A. G. Simmonds (Eds.), Academic, New York, 1971.
13. A. Cammarata and K. S. Rogers, The Interpretation of Drug Action through Linear Free Energy Relationships, in *Advances in Linear Free Energy Relationships*, N. R. Chapman and J. Shorter (Eds.), Plenum Press, New York, 1972.
14. A. Verloop, The Use of Linear Free Energy Parameters and Other Experimental Constants in Structure–Activity Studies, in *Drug Design*, Vol. 3, E. J. Ariens (Ed.), Academic, New York, 1972.
15. W. J. Dunn, Quantitative Structure–Activity Relationships, in *Annual Reports in Medicinal Chemistry*, Vol. 8, R. V. Heinzelman (Ed.), Academic, New York, 1973.
16. P. J. Goodford, Prediction of Pharmacological Activity by the Method of Physicochemical–Activity Relationships, in *Advances in Pharmacology and Chemotherapy*, Vol. 11, S. Garranttini et al. (Eds.), Academic, New York, 1973.
17. C. Hansch, Quantitative Approaches to Pharmacological Structure–Activity Relationships, in *Structure–Activity Relationships*, C. J. Cavallito (Ed.), Pergamon, Oxford, 1973.
18. G. Redl, R. D. Cramer, and C. E. Berkoff, Quantitative Drug Design, *Chem. Soc. Rev.*, **3**, 273, (1974).

References

19. C. Hansch, Enzyme Study as a source of strategy in Drug Design, *Adv. Pharmacol. Chemother.*, **13**, 45 (1975).
20. S. Wold and M. Sjostrom, Linear Free Energy Relationships as Tools for Investigating Chemical Similarity—Theory and Practice in *Advances in Linear Free Energy Relationships*, Vol. 2, N. B. Chapman and J. Shorter (Eds.) Plenum Press, New York, in press.
21. R. D. Cramer, Quantitative Drug Design, in *Annual Reports in Medicinal Chemistry*, Vol. 11, F. H. Clarke (Ed.), Academic, New York, 1976.
22. C. Hansch, Recent Advances in Biochemical QSAR, in *Advances in Linear Free Energy Relationships*, Vol. 2, N. R. Chapman and J. Shorter (Eds.), Plenum Press, New York, in press.
23. Y. C. Martin, Advances in the Methodology of Quantitative Drug Design, in *Drug Design*, Vol. VIII, E. J. Ariens (Ed.), Academic, New York, 1978.
24. C. Hansch and T. Fujita, ρ-σ-π Analysis. A Method for the Correlation of Biological Activity and Chemical Structure, *J. Am. Chem. Soc.*, **86**, 1616 (1964).
25. L. P. Hammett, *Physical Organic Chemistry*, McGraw Hill, New York, 1940.
26. A. Leo, C. Hansch, and D. Elkins, Partition Coefficients and Their Uses, *Chem. Rev.*, **71**, 525 (1971).
27. R. N. Smith, C. Hansch, and M. A. Ames, Selection of a Reference Partitioning System for Drug Design Work, *J. Pharm. Sci.*, **64**, 599 (1975).
28. G. G. Nys and R. F. Rekker, Statistical Analysis of a Series of Partition Coefficients with Special Reference to the Predictability of Folding of Drug Molecules. The Introduction of Hydrophobic Fragmental Constants (F-Values). *Eur. J. Med. Chem.*, **9**, 521 (1973).
29. G. G. Nys and R. F. Rekker, The Concept of Hydrophobic Fragmental Constants (F-Values). II. Extension of Its Applicability to the Calculation of Aromatic and Heteroaromatic Structures., *Eur. J. Med. Chem.*, **9**, 361–375 (1974).
30. R. F. Rekker, *The Hydrophobic Fragmental Constant*, Elsevier, Amsterdam, 1977.
31. A. Leo, P. Y. C. Jow, C. Silipo, and C. Hansch, Calculation of Hydrophobic Constant (log P) from π and F Constants, *J. Med. Chem.*, **18**, 865 (1975).
32. L. H. M. Janssen and J. H. Perrin, Some Theoretical Observations on the Estimation of Partition Coefficients from π and f constants, *Eur. J. Med. Chem.*, **11**, 197 (1976).
33. A. J. Hopfinger and R. D. Battershell, Application of SCAP to Drug Design. 1. Prediction of Octanol–Water Partition Coefficients Using Solvent-Dependent Conformational Analysis, *J. Med. Chem.*, **19**, 569 (1976).
34. J. T. Penniston, L. Beckett, D. L. Bentley, and C. Hansch, Passive Permeation of Organic Compounds through Biological Tissue: a Non-Steady-State Theory, *Mol. Pharmacol.*, **5**, 333 (1969).
35. J. C. Dearden and M. S. Townend, Digital Computer Simulation of the Drug Transport Process, paper presented at the Symposium on Chemical Structure–Biological Activity Relationships Quantitative Approaches, Suhl, G. D. R., October 1976.
36. J. W. McFarland, On the Parabolic Relationship between Drug Potency and Hydrophobicity, *J. Med. Chem.*, **13**, 1192 (1970).
37. G. L. Biagi, A. M. Barbaro, O. Gandolfi, M. C. Guerra, and G. Cantelli-Forti, R_m Values of Steroids as an Expression of Their Lipophilic Character in Structure–Activity Studies, *J. Med. Chem.*, **18**, 873 (1975).
38. D. Brown and D. Wordcock, Relationships between Hansch's π parameters and R_m Values Determined on Polyamide Thin Layers, *J. Chromatogr.*, **105**, 33 (1975).

39. R. M. Carlson, R. E. Carlson, and H. L. Kopperman, Determination of Partition Coefficients by Liquid Chromatography, *J. Chromatogr.*, **107**, 219 (1975).
40. J. M. McCall, Liquid–Liquid Partition Coefficients by High-Pressure Liquid Chromatography, *J. Med. Chem.*, **18**, 549 (1975).
41. C. G. Swain and E. C. Lupton, Field and Resonance Components of Substituent Effects, *J. Am. Chem. Soc.*, **90**, 4328 (1968).
42. C. Hansch, A. Leo, S. H. Unger, K. H. Kim, D. Nikaitani, and E. J. Lien, "Aromatic" Substituent Constants for Structure–Activity Correlations, *J. Med. Chem.*, **16**, 1207 (1973).
43. R. W. Taft, Jr., Separation of Polar, Steric, and Resonance Effects in Reactivity, in *Steric Effects in Organic Chemistry*, M. S. Newman (Ed.), Wiley, New York, 1956.
44. S. H. Unger and C. Hansch, Quantitative Models of Steric Effects, in *Progress in Physical Organic Chemistry*, Vol. 12, A. Streitwieser, Jr., and R. W. Taft (Eds.), Wiley–Interscience, New York, 1976.
45. A. Verloop, W. Hoogenstraaten, and J. Tipker, Development and Application of New Steric Substituent Parameters in Drug Design, in *Drug Design*, Vol. VII, E. J. Ariens (Ed.), Academic, New York, 1976.
46. Z. Simon, Specific Interactions, Intermolecular Forces, Steric Requirements, and Molecular Size, *Angew. Chem.*, **13**, 719 (1974).
47. C. Silipo and C. Hansch, Correlation Analysis. Its Application to the Structure–Activity Relationship of Triazines Inhibiting Dihydrofolate Reductase, *J. Am. Chem. Soc.*, **97**, 6849 (1975).
48. G. W. Snedecor and W. C. Cochran, *Statistical Methods*, 6th ed., The Iowa State University Press, Ames, Iowa, 1967.
49. N. R. Draper and H. Smith, *Applied Regression Analysis*, Wiley, New York, 1966.
50. P. J. Lewi, Computer Technology in Drug Design, in *Drug Design*, Vol. VII, E. J. Ariens (Ed.), Academic, New York, 1976.
51. P. N. Craig, C. Hansch, J. W. MacFarland, Y. C. Martin, W. P. Purcell, R. Zahradnik, Minimal Statistical Data for Structure–Function Correlations, *J. Med. Chem.*, **14**, 447 (1971).
52. W. H. Davis, Jr., and W. A. Pryor, Measures of Goodness of Fit in Linear Free Energy Relationships, *J. Chem. Ed.*, **53**, 285 (1976).
53. W. J. Dixon (Ed.), *BMD—Biomedical Computer Programs*, 3rd ed., University of California Press, Berkeley, CA, 1973.
54. G. M. Furnival and R. W. Wilson, Jr., Regressions by Leaps and Bounds, *Technometrics*, **16**, 499 (1974).
55. A. E. Hoerl and R. W. Kennard, Ridge Regression: Biased Estimation for Nonorthogonal Problems, *Technometrics*, **12**, 55 (1970).
56. R. R. Hocking, The Analysis and Selection of Variables in Linear Regression, *Biometrics*, **32**, 1 (1976).
57. P. J. Lewi, The Use of Multivariate Statistics in Industrial Pharmacology, in *Encyclopedia of Pharmacology and Therapeutics*, 1977.
58. P. N. Craig, Interdependence between Physical Properties and Selection of Substituent Groups for Correlation Studies, *J. Med. Chem.*, **14**, 680 (1971).
59. J. G. Topliss and R. J. Costello, Chance Correlations in Structure–Activity Studies Using Multiple Regression Analysis, *J. Med. Chem.*, **15**, 1066 (1972).

References

60. S. H. Unger and C. Hansch, On Model Building in Structure–Activity Relationships. A Reexamination of Adrenergic Blocking Activity of β-Halo-β-arylalkylamines, *J. Med. Chem.*, **16**, 745 (1973).
61. S. M. Free and J. W. Wilson, A Mathematical Contribution to Structure–Activity Studies, *J. Med. Chem.*, **7**, 395 (1964).
62. A. Cammarata, Interrelationship of the Regression Models Used for Structure–Activity Analysis, *J. Med. Chem.*, **15**, 573 (1972).
63. T. Fujita and T. Ban, Structure–Activity Study of Phenethylamines as Substrates of Biosynthetic Enzymes of Sympathetic Transmitters, *J. Med. Chem.*, **14**, 148 (1971).
64. H. Kubinyi and O. H. Kehrhahn, Quantitative Structure–Activity Relationships. 3. A Comparison of Different Free-Wilson Models, *J. Med. Chem.*, **19**, 1040 (1976).
65. H. Kubinyi and O. H. Kehrhahn, Quantitative Structure–Activity Relationships. 1. The Modified Free-Wilson Approach, *J. Med. Chem.*, **19**, 579 (1976).
66. H. Kubinyi, Quantitative Structure–Activity Relationships. 2. A Mixed Approach Based on Hansch and Free-Wilson Analysis, *J. Med. Chem.*, **19**, 587 (1976).
67. P. N. Craig, Comparison of the Hansch and Free-Wilson Approaches to Structure–Activity Correlations, in *Biological Correlations—The Hansch Approach*, Advances in Chemistry Series, No. 114, American Chemical Society, Washington, D.C., 1972.
68. L. B. Kier, *Molecular Orbital Theory in Drug Research*, Academic, New York, 1971.
69. W. B. Neely, The Use of Molecular Orbital Theory in Pharmacological Studies, in *A Guide to Molecular Pharmacology–Toxocology*, Part II, R. M. Featherstone (Ed.), Dekker, New York, 1973.
70. J. P. Green, C. L. Johnson, and S. Kang, Application of Quantum Chemistry to Drugs and Their Interactions, *Annu. Rev. Pharm.*, **14**, 319 (1974).
71. J. J. Kaufman and W. S. Koski, Physicochemical, Quantum Chemical, and Other Theoretical Techniques for the Understanding of the Mechanism of Action of CNS Agents: Psychoactive Drugs, Narcotics, and Narcotic Antagonists and Anesthetics, in *Drug Design*, Vol. V, E. J. Ariens (Ed.), Academic, New York, 1975.
72. W. G. Richards and M. E. Black, Quantum Chemistry in Drug Research, in *Progress in Medicinal Chemistry*, Vol. 11, G. P. Ellis and G. B. West (Eds.), American Elsevier, New York, 1975.
73. R. Christoffersen, Use of Quantum Chemistry in Development and Analysis of Anticancer Drugs, *Cancer Chemother. Rep.*, Part 2, **4(4)**, 47 (1974).
74. R. E. Christoffersen, Molecules of Pharmacological Interest, in *Quantum Mechanics of Molecular Conformations*, B. Pullman (Ed.), Wiley, New York, 1976.
75. R. E. Christoffersen and R. P. Angeli, Quantum Pharmacology, in *The New World of Quantum Chemistry*, B. Pullman and R. Parr (Eds.), D. Reidel Publ. Co., Dordrecht, Holland, 1976.
76. J. I. Fernandez-Alonso, Organic Molecules. Studies by Semi-empirical Methods, in *Quantum Mechanics of Molecular Conformations*, B. Pullman (Ed.), Wiley, New York, 1976.
77. L. B. Kier, Receptor Mapping Using Molecular Orbital Theory, in *Fundamental Concepts in Drug–Receptor Interactions*, J. F. Danielli, J. F. Morgan, and D. J. Triggle (Eds.), Academic, New York, 1970.
78. L. B. Kier, Molecular Orbital Studies of Biological Molecule Conformations, in *Biological Correlations—The Hansch Approach*, Advances in Chemistry Series, No. 114, American Chemical Society, Washington, D.C., 1972.

79. G. L. Kirschner and B. R. Kowalski, The Application of Pattern Recognition to Drug Design, in *Drug Design*, Vol. VIII, E. J. Ariens (Ed.), Academic, New York, 1978.
80. K. L. Ting, R. C. T. Lee, G. W. A. Milne, M. Shapiro, and A. M. Guarino, The Applications of Artificial Intelligence: Relationships between Mass Spectra and Pharmacological Activity of Drugs, *Science*, **180**, 417 (1973).
81. J. T. Clerc, P. Naegeli, and J. Seibl, Artificial Intelligence, *Chimia*, **27**, 639 (1973).
82. C. L. Perrin, Testing of Computer-Assisted Methods for Classification of Pharmacological Activity, *Science*, **183**, 551 (1974).
83. H. Abe, S. Kumazawa, T. Taji, and S. Sasaki, Applications of Computerized Pattern Recognition: A Survey of Correlations between Pharmacological Activities and Mass Spectra, *Biomed. Mass Spectrosc.*, **3**, 151 (1976).
84. S. A. Hiller, Y. E. Golender, A. B. Rosenblit, L. A. Rastrigin, A. B. Glaz, Cybernetic Methods of Drug Design. I. Statement of the Problem—The Perceptron Approach, *Comp. Biomed. Res.*, **6**, 411 (1973).
85. Y. C. Martin, J. B. Holland, C. H. Jarboe, and N. Plotnikoff, Discriminant Analysis of the Relationship between physical Properties and the Inhibition of Monoamine Oxidase by Aminotetralins and Aminoindans, *J. Med. Chem.*, **17**, 409 (1974).
86. K. C. Chu, Applications of Artificial Intelligence to Chemistry. Use of Pattern Recognition and Cluster Analysis to Determine the Pharmacological Activity of Some Organic Compounds, *Anal. Chem.*, **46**, 1181 (1974).
87. B. R. Kowalski and C. F. Bender, The Application of Pattern Recognition to Screening Prospective Anticancer Drugs. Adenocarcinoma 755 Biological Activity Test, *J. Am. Chem. Soc.*, **96**, 916 (1974).
88. R. J. Mathews, A Comment on Structure–Activity Correlations Obtained Using Pattern Recognition Methods, *J. Am. Chem. Soc.*, **97**, 935 (1975).
89. S. H. Unger, Discussion of Pattern Recognition, *Cancer Chemother. Rep.*, Part 2, **4**, 45 (1974).
90. K. C. Chu, R. J. Feldman, M. B. Shapiro, G. F. Hazard, Jr., and R. I. Geran, Pattern Recognition and Structure–Activity Relationship Studies. Computer-Assisted Prediction of Antitumor Activity in Structurally Diverse Drugs in an Experimental Mouse Brain System, *J. Med. Chem.*, **18**, 539 (1975).
91. A. J. Stuper and P. C. Jurs, Classification of Psychotropic Drugs as Sedatives or Tranquilizers Using Pattern Recognition Techniques, *J. Am. Chem. Soc.*, **97**, 182 (1975).
92. A. Cammarata and G. K. Menon, Pattern Recognition. Classification of Therapeutic Agents According to Pharmacophores, *J. Med. Chem.*, **19**, 739 (1976).
93. G. K. Menon and A. Cammarata, Pattern Recognition II. Investigation of Structure–Activity Relationships, *J. Pharm. Sci.*, **66**, 304 (1977).
94. A. J. Stuper and P. C. Jurs, ADAPT: A Computer System for Automated Data Analysis Using Pattern Recognition Techniques, *J. Chem. Infor. Comp. Sci.*, **16**, 99 (1976).
95. A. J. Stuper, W. E. Brugger, P. C. Jurs, A Computer System for Structure–Activity Studies Using Chemical Structure Information Handling and Pattern Recognition Techniques, in *Chemometrics: Theory and Applications*, B. R. Kowlaski (Ed.), American Chemistry Society Symposium Series, No. 52, American Chemical Society, Washington, D.C., 1977.
96. A. J. Stuper and P. C. Jurs, Structure Activity Studies of Barbiturates Using Pattern Recognition Techniques, *J. Pharm. Sci.*, 167, 745 (1978).
97. R. D. Cramer III, G. Redl, and C. E. Berkoff, Substructural Analysis. A Novel Approach to the Problem of Drug Design, *J. Med. Chem.*, **17**, 533 (1974).

References

98. G. W. Adamson and J. A. Bush, Method for Relating the Structure and Properties of Chemical Compounds, *Nature*, **248**, 406 (1974).
99. G. W. Adamson and J. A. Bush, A Comparison of the Performance of Some Similarity and Dissimilarity Measures in the Automatic Classification of Chemical Structures, *J. Chem. Inf. Comp. Sci.*, **15**, 55 (1975).
100. G. W. Adamson and J. A. Bush, Evaluation of an Empirical Structure–Activity Relationship for Property Prediction in a Structurally Diverse Group of Local Anaesthetics, *J. Chem. Soc. Perkin I*, No. 2, **168** (1976).
101. W. E. Brugger, A. J. Stuper, and P. C. Jurs, Generation of Descriptors from Molecular Structures, *J. Chem. Inf. Comp. Sci.*, **16**, 105 (1976).
102. D. S. Dierdorf and B. R. Kowalski, Three Dimensional Molecular Structure–Biological Activity Correlations by Pattern Recognition, NTIS Report No. AD-785863/2GA, 1974.
103. L. J. Soltzberg and C. L. Wilkins, Computer Recognition of Activity Class from Molecular Transforms, *J. Am. Chem. Soc.*, **98**, 4006 (1976).
104. L. J. Soltzberg and C. L. Wilkins, Molecular Transforms: A Potential Tool for Structure–Activity Studies, *J. Am. Chem. Soc.*, **99**, 439 (1977).
105. P. Gund, Three-Dimensional Pharmacophoric Pattern Searching, *Mol. Subcell. Biol.*, **5**, 117 (1977).
106. J. R. McGill and B. R. Kowalski, Intrinsic Dimensionality of Smell, *Anal. Chem.*, **49**, 596 (1977).
107. W. E. Brugger and P. C. Jurs, Extraction of Important Molecular Features of Musk Compounds Using Pattern Recognition Techniques, *J. Agr. Food Chem.*, **25**, 1158 (1977).
108. C. Hansch, S. H. Unger, and A. B. Forsythe, Strategy in Drug Design. Cluster Analysis as an Aid in the Selection of Substituents, *J. Med. Chem.*, **16**, 1217 (1973).
109. R. F. White and T. M. Lewinson, Probabilistic Clustering for Attributes of Mixed Type with Biopharmaceutical Applications, *J. Am. Stat. Assoc.*, **72**, 271 (1977).
110. L. Hodes, G. F. Hazard, R. I. Geran, and S. Richman, A Statistical-Heuristic Method for Automated Selection of Drugs for Screening, *J. Med. Chem.*, **20**, 469 (1977).

CHAPTER 2
Pattern Recognition Principles

Implicit in the design of new drugs is the assumption that structurally similar compounds possess similar biological activities. Developing definitions of this similarity has proved to be an extremely complex task, as evidenced by the variety of parameters used in attempts at developing empirical equations relating drug structure to biological activity. By far the most popular method of developing such relations has been through the use of regression analysis. The purpose of this approach is the development of empirical equations that relate various combinations of physical, chemical, or structural parameters to the biological response of the compound. These methods offer attractive capabilities for studies involving a small to medium number of homologues.

Often it is desired to develop structure–activity relations for a medium to large number of structurally diverse compounds. The assumptions involved in the use of regression analysis on homologues cannot readily be extended to structurally diverse sets of compounds. For such cases, alternative methods are needed for classifying the compounds so that a general statement concerning their action can be made. Such operations are particularly relevant to applications concerned with determining which few of a large number of compounds are most likely to exhibit a particular effect.

The approach normally taken in choosing which compounds are most promising is intuitive, that is, an educated guess is made. Use of the educated guess is quite valid. The scientist who has worked with a particular class of compounds often develops an intuitive insight into what features a compound must possess to be of possible utility. However, the success of the intuitive approach is inversely proportional to the number of parameters involved and the degree of their interdependence. Our intuitive outlook could certainly benefit from an organized approach to developing these relationships. The techniques of pattern recognition offer an aid in developing such rules.

Numerous texts on the subject of pattern recognition exist (1–17). This is no doubt a reflection of the wide number of areas that employ these methods. Applications of pattern recognition to chemical problems first appeared in the mid 1960s (18,19) with studies of mass spectra. Since then papers have described work in a variety of areas. Works prior to 1975 have been cited in reviews (20–23). More recently, papers have been published describing applications in mass spectrometry (24–26), infrared spectroscopy (27,28), nuclear magnetic spectrometry (29–32), stationary electrode polarography (33,34), materials science and mixture analysis (35–37), and the modeling of chemical experiments (38–44). The success achieved in these areas using pattern recognition methods as aids in the empirical rule making process leads us to believe that these methods would prove equally useful in the development of structure–activity relations.

Pattern recognition methods are uniquely suited to a variety of studies because of several novel attributes. No exact functional form is fitted to the data; rather, relationships are sought that provide a definition of similarity among diverse groups of data. In essence, pattern recognition techniques can be thought of as providing relations that uncover common properties. Once such relations are developed they may be used to infer the properties of members that were not part of the original data set.

One interesting feature of these techniques is their ability to deal with high dimensional data (data for which more than three measurements are used to represent each object). In addition, these techniques can deal with multisource data or data in which the relationships are discontinuous. When properly used, pattern recognition techniques allow the scientist to develop criteria that relate general properties to a subset of the total number of measurements. Once the important measurements are identified, they can be used to guide the development of subsequent experiments. For example, if one were to find that 10 structural parameters were useful in classifying compounds as exhibiting or not exhibiting a particular effect, then one might hypothesize several as yet unstudied structures and use the results from the pattern recognition analysis to make an educated guess concerning the likelihood that those structures will exhibit this effect. Alternatively, the fact that the particular 10 parameters were shown to be important may lead to added insights into the problem. This ability to pick a subset of the original measurements that contains the bulk of the total information content is what affords pattern recognition techniques their utility in a wide variety of fields.

BASIC PATTERN RECOGNITION METHODS

To begin our discussion of pattern recognition methods we have to define what is meant by classification of an object or set of objects. Classification is the act of developing a rule that categorizes a group of objects, as opposed to recognition, which is the use of the classification rule to place an unknown into any one of several possible categories. The classification rule hopefully

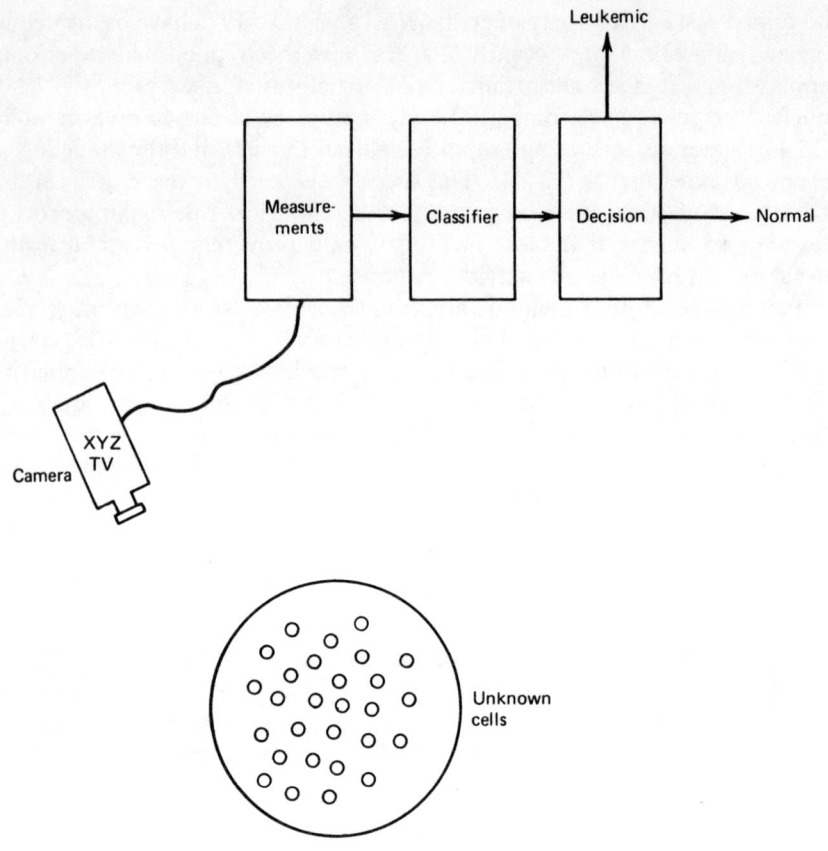

Figure 2.1 Example of an optical pattern recognition system.

will account for the properties held by a large set of samples. If this is the case, applying that rule to an unknown results in the unknown being properly classified. The reliability of the recognition process is a complex function of the manner in which the classification rule was developed and the amount and type of information used to develop the rule. This process is somewhat analogous to the way humans approach a problem; that is, a classification rule is created in the form of a hypothesis developed from a sampling of data. To test the hypothesis, one applies the rule to members not in the set of data used to develop the original hypothesis. If the hypothesis is correct, then it will prove true for data not in the original observation. If exceptions are found, the hypothesis is discarded as incorrect and a new one is sought. Of

Basic Pattern Recognition Methods 33

course the data set itself defines the limits of any classification rule. Only those elements that are in the domain of the data used to develop the rule can be properly classified by that rule.

The development of a classification rule and its use in recognition involves more than the classification process. Therefore, a simple example is given here to demonstrate the overall problem. Once the series of steps necessary to effect the procedure are defined, the specific properties and requirements of each part of the system will be expanded upon.

As an example of how classification rules can be developed let us consider the following imaginary and somewhat fanciful problem. Suppose a clinical laboratory wishes to automate its processing such that a machine performs the identification of the various types of abnormal blood cells. As a pilot project it is decided to try to distinguish leukemic cells from normal cells through the use of an optical sensing system. Such a system might be as shown in Figure 2.1. The camera provides optical measurements concerning the various types of blood cells. These measurements or features are given to a device that chooses those most relevant. The chosen features are then passed on to a classifier that develops a rule to classify a set of prototypes and this rule is then used to make a decision on the type of blood cells presented to the classifier.

Let us now consider how such a feature extractor and classifier could be designed. Since it is possible to take a large number of optical measurements, guidance is needed in selecting those most useful. Let us assume that the designer has noted that one feature is that leukemic cells are lighter in color than normal cells. Thus it seems we might initially attempt to classify the cells on the basis of brightness alone. If the brightness exceeds some limit, say X_0, then it is classified as normal. To investigate what this limit might be, a number of cells of each type are selected and a histogram plot is constructed as shown in Figure 2.2. This histogram indeed shows that leukemic cells are generally lighter than normal cells; however, there is no one value of brightness, X_0, that will assure us of a reasonable number of misclassifications.

Since the current level of reliability is too low, we must seek other features that would be useful in differentiating the various cell types. Suppose leukemic cells have a more prominent cell structure than normal cells. The camera could then be used to measure light–dark transitions and obtain a measure of texture for each of the prototype cells in the test set. This results in a scatter plot such as that for Figure 2.3. Note that this is equivalent to treating each of our measurements as components of a vector such that $X = [x_1, x_2]$. The net result is a distribution of points in space such that each point is represented by its corresponding vector. Thus the following rule might be conceived: If the vector is above line AB, classify the cell as normal; if it is below the line, classify as leukemic.

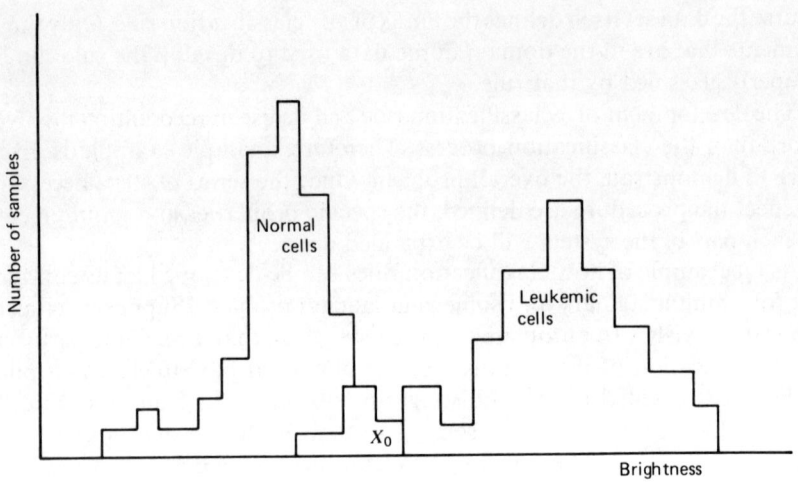

Figure 2.2 A possible histogram of brightness measurements.

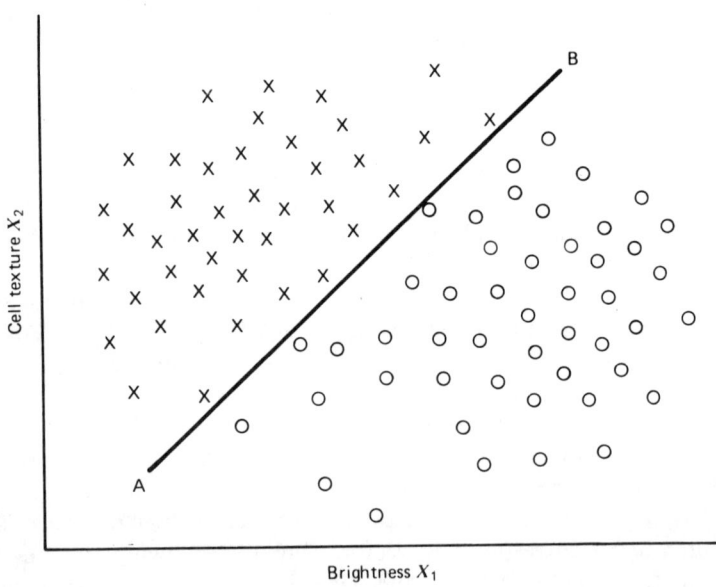

Figure 2.3 Scatter of cell texture versus brightness. (×) Normal cells; (○) leukemic cells.

Of course we have merely classified our test set. It would seem that we should test how well we can recognize other leukemic or normal cells by obtaining more samples and using our system to see if they are classified correctly. If our test set was sufficiently large, then we would expect that it contained a representative sampling of all possible normal and leukemic cell types and that our current rule was quite general. However, at this point it is not clear how to estimate the reliability of our rule using the test set alone. Additionally, we have no method of dealing with other types of abnormal cells. Clearly, the rule we have developed would not be able to differentiate leukemic or normal from other cell types, since examples of these were never included in the original classification test set. Indeed, to proceed further, we would need a generalized approach to the concept that yielded our original classification rule. Such an approach is shown in Figure 2.4.

Figure 2.4 shows a flow diagram of the basic processes that must be performed to apply pattern recognition to any given problem. System refers to the collection of objects that are to undergo analysis. It is hoped that by making a series of measurements on the objects in the system a rule can be developed that will be useful in generalizing the properties of the system. An example of a chemical rule is the answer to the question, "Does this compound contain a carbonyl functional group?" Obviously, the property of containing a carbonyl group is a measureable parameter. In cases such as this the rule can be expressed as a function whose value is positive if the compound has an absorbance at 1700 cm^{-1} or if it undergoes certain characteristic reactions (45), and negative otherwise. A positive value for the equation is equated with a positive answer to the question. However, if several measurements are required to develop such a rule, it becomes necessary to use deductive aids such as pattern recognition techniques to develop the function. This is especially the case when the measurements are complex or interrelated.

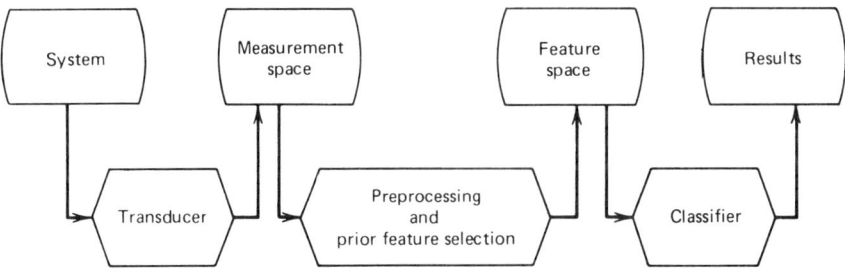

Figure 2.4 Flow diagram of a basic pattern recognition system.

The device that performs measurements on the system is called a transducer. An extensive discussion of transducers relevant to structure–activity studies is given in Chapter 3.

There are many types of chemical transducers, for example, mass spectrometers, photometers, NMR, GC, and viscometers, that can be used to provide chemically relevant measurements. These measurements are called descriptors.

The result of making measurements on a member of the system can be expressed as a vector such that each of its components is a quantity descriptive of that measurement, that is, the vector

$$\mathbf{X} = \begin{bmatrix} x_1 \\ x_2 \\ x_3 \\ \vdots \\ x_N \end{bmatrix}$$

is the representation of a member having N measurements made on it. The results of making measurements on the entire system can then be expressed in the form of a matrix. For a system of M members each with N measurements, the matrix

$$\begin{bmatrix} x_{11} & x_{21} & x_{31} & \cdots & x_{M1} \\ x_{12} & x_{22} & x_{32} & \cdots & x_{M2} \\ x_{13} & x_{23} & x_{33} & \cdots & x_{M3} \\ \vdots & \vdots & \vdots & & \vdots \\ x_{1N} & x_{2N} & x_{3N} & & x_{MN} \end{bmatrix}$$

results. This matrix is one representation of the measurement space. It is often convenient to picture this matrix geometrically, in which case each member is a point in the N dimensional hyperspace. Thus the measurements provided by the transducer are responsible for the distribution of the data in the space. Clearly, there are two types of measurement spaces to be found, those resulting from single source measurements and those from multisource measurements.

Single source measurements originate from the same transducer. In these cases the components of the measurement vector are of the same scale. A mass spectrometer is an example of a single source measurement. Each component of the data vector represents the ion intensity at a particular mass unit.

In multisource data, each measurement can be the result of an independent transducer, and each can have a different scale, origin, distribution, and so on from all the other measurements. Therefore, there need be no direct functional relationship between the measurements from multisource data as there must be, for example, in an absorbance versus concentration plot.

For many chemical problems, and especially for those employing multisource data, it is difficult to know in advance which features of the measurement space will yield a satisfactory solution. The generation of sufficiently informative multisource measurements can become in itself a major part of a pattern recognition system.

Since chemical systems are very complex, there is often a tendency to overdetermine the data set. In such cases different measurements may often contain redundant information. The process by which the most useful features are selected from the measurement space to be used in the pattern recognition analysis is known as feature selection. The particular technique used to select features from the measurement space depends on the criteria used to define useful or important features. Many transforms are available that select features or combinations of features to maximize clustering, minimize divergence, or account for the maximum amount of variance with the fewest number of features. Generally the object is to "do the most with the least"; the fewer the number of features that must be included the easier the classification task and, in some cases, the more reliable the results. The net result of feature selection is the culling out from the measurement space of a subset of descriptors that can be used to perform the operation known as classification.

Classification is the operation that groups the members of the system into one of several classes, based on the description as it exists in feature space. It naturally arises from a geometrical interpretation of feature space. If the measurements contain information about the data, then the assumption is that similar data will tend to cluster in definable, limited, regions of the N-dimensional space. Generally a subset of the total data set is taken and the rule is developed using this "training set." The utility of this rule can then be tested using the members of the data set not included in the training set.

The result of these operations is a rule that, hopefully, is able to correctly classify all or most of the members in the system. If the set of data used to develop the system is truly representative of all the possible members that could have been included in the system, then the rule will very likely hold for elements originally excluded from the system.

It should now be clear that pattern recognition is performed through application of the three operations: transduction, preprocessing, and classification. These operations create or provide a measurement space, a feature space, and a rule that allows the division of the system into classes. While we have divided these operations into separate parts, this division is, in a sense, artificial in that certain of the techniques used to effect one of the operations may have utility in the other as well. However, these divisions do form a set of concepts that aid in the discussion of the various techniques used to implement them.

The next few sections contain discussions of the basis for techniques employed in effecting the operations of preprocessing and certain types of classification. Transduction is a highly problem oriented function and is dealt with in a separate chapter.

PREPROCESSING

Preprocessing methods alter the form of the output from a transducer. These methods include scaling, normalization, clustering transformations, feature selection, multidimensional scaling, and nonlinear mapping.

Scaling and Normalization

Scaling and normalization are required to convert the units from different transducers to a compatible form. This is necessary when dealing with multisource data, as the units may differ by several orders of magnitude. In such cases the larger valued descriptors tend to dominate those of smaller value. One preprocessing method that is quite useful in correcting this imbalance is autoscaling.

This method simultaneously scales and normalizes the data by translating it so that the mean is zero and normalizing it so that each descriptor has a standard deviation of S. The equations describing this process are given below.

$$X'_{ij} = \frac{S(X_{ij} - \bar{X}_j)}{\sigma_j} \quad (2.1)$$

$$\bar{X}_j = \frac{1}{N} \sum_{i=1}^{N} X_{ij} \quad (2.2)$$

$$\sigma_j^2 = \frac{1}{N} \sum_{i=1}^{N} (X_{ij} - \bar{X}_j)^2 \quad (2.3)$$

The X_{ij} are the elements of the matrix representation of the data formed from the column vectors representing each data point, and the X'_{ij} are the elements of the data matrix resulting from the scaling operation. Autoscaling can be looked upon as "squaring up the space," since it places the data inside a hypercube. Although autoscaling alters the spread of the data, it does not alter the number of features or the basic geometry of the clustering. Adding new members to a large data set often requires only the use of equation 2.1, as $\bar{X}_j$ and σ_j are not normally significantly altered.

Once the data have been autoscaled, it may be desirable to further normalize the space so that measurements contributing most to the clustering of

Preprocessing

the data are weighted more than those contributing least. One of the simplest methods of achieving such a transform is through the use of variance weighting. This transform attempts to form a new space $\mathbf{X}'$, from the old space $\mathbf{X}$ using the transform

$$\mathbf{X}'_{ik} = Q_k \mathbf{X}_{ik} \quad (2.4)$$

$$Q_k = \frac{\sum_i^L \sum_{j=i}^L P_i P_j X_{ij}^k}{\sum_i^L P_i X_{ii}^k} \quad (2.5)$$

$$X_{ij}^k = \sum_{l,m} (\mathbf{X}_{lk}^i - \mathbf{X}_{mk}^j)^2 \quad (2.6)$$

L is the number of classes, P_i is the simple probability of a point being in class i (number in class/number in data set), X_{ij}^k is the interclass variance found from all two-point combinations not in the same class. X_{lk}^i is the kth variable for the lth member of class i, X_{ii}^k is the intraclass variance found from all two-point combinations within the same class, and Q_k is the weighting factor formed for the kth variable in the data set. The larger the value of the Q_k the more important the variable is for the classification.

The major drawback to a transform of this type is that it attempts to weight the best individual variables the most. It is well known that the best variables identified one at a time will not necessarily form the best set of variables. Thus the utility of the ranking provided by this measurement should be viewed with some skepticism. As with all the techniques the applicability of the method depends on the nature of the problem being addressed.

Clustering Transformations

Although operations such as autoscaling can correct for effects due to the differing units of multisource data and variance weighting attempts to weight the features according to their ability to separate, such operations change all members of a data set equally. Often it is necessary to preprocess the data such that a single class is favored. These types of operations are used to minimize the interclass distance for some particular class in hopes of tightening that classes clustering and thus making the classification task easier. Such a transformation can be effected by weighting the components of the feature space according to their importance in the clustering of the class of interest. One method of effecting such a transformation is to generate a linear transform,

$$\mathbf{X}^* = \mathbf{W}\mathbf{X} \quad (2.7)$$

such that after transforming from $\mathbf{X}$ space to $\mathbf{X}^*$ space the interclass distance for the class of interest is minimized.

The derivation of this transformation is relatively straightforward and has been presented in detail elsewhere (14). We present here an extremely simplified explanation that will allow its application as a preprocessing aid.

The intraset distance for a set of pattern points $\{X^i, i = 1, 2, 3, \ldots, K\}$ is given by

$$\bar{D}^2 = 2 \sum_{k=1}^{N} \sigma_k^2 \tag{2.8}$$

where N is the dimensionality of the space and σ_k^2 is the unbiased variance of the class calculated from the relationship

$$\sigma_k^2 = \frac{1}{K-1} \sum_{i=1}^{K} (a_k^i - \bar{a}_k^i)^2 \tag{2.9}$$

$\bar{a}_k^i$ is the average value for each dimension of the pattern vector calculated from the relation

$$\bar{a}_k^i = \frac{1}{K} \sum_{i=1}^{K} a_k^i \tag{2.10}$$

K is the number of members in the class whose interclass distance is being minimized. What we wish to find is the transform matrix $\mathbf{W}$, which minimizes the interest distance for a particular class in the new space. Since we are only interested in scale factor changes we may force $\mathbf{W}$ to be a diagonal matrix. The distance in the new space would then be

$$\bar{D}^2 = 2 \sum_{k=1}^{N} (w_{kk} \sigma_k^2) \tag{2.11}$$

There are two obvious ways of constraining these minimization procedures:

Case 1

$$\sum_{k=1}^{N} w_{kk} = 1$$

Case 2

$$\prod_{k=1}^{N} w_{kk} = 1$$

For case 1 the transform matrix is calculated from

$$w_{kk} = \frac{1}{\sigma_k^2 \sum_{j=1}^{N} (1/\sigma_j^2)} \tag{2.12}$$

Here we note that if σ_k^2 is large w_{kk} is small. This appears intuitively satisfying, since those dimensions that have the greatest variance would contribute

Preprocessing

least to the clustering of the data set and thus seem to be weighted appropriately. Conversely, if σ_k^2 is small its contribution to the clustering is large and we would wish to weight these dimensions more highly.

For case 2 the transform matrix is calculated using

$$w_{kk} = \frac{1}{\sigma_k} \left(\prod_{j=1}^{N} \sigma_j \right)^{1/N} \tag{2.13}$$

Here we note that the feature weighting coefficient is inversely proportional to the standard deviation of the kth measurement. Again, those measurements with the least variance contribute most to the clustering and are therefore weighted the most.

Clearly, these transforms are straightforward to apply using only simple calculations to construct the transform. Since the transform matrix is diagonal the number of matrix multiplication operations is drastically reduced resulting in a large savings in time. In addition, since the transform is linear it can be applied very easily to preprocess new members that are to be used in a classification or recognition process. Of course minimization of interclass distance for several different classes requires separate minimizations for each class since the transform is effective for only one class in each application. Also, since the new space is a linear combination of the original space the relationship between the new measurements and the old ones is somewhat blurred. As with every preprocessing technique the applicability of the clustering transform depends on the problem being addressed.

Feature Selection

One disadvantage of preprocessing methods that operate on all the descriptors is that they include measurements that are not germane to the classification problem. This can result in highly nonoptimal situations, especially since useless measurements tend to increase the error in the classification process, as well as the complexity and cost of these operations. Since not all the measurements made on a system are meaningful for every problem, methods that reduce the number of these descriptors are necessary. These methods come under the heading of feature selection.

Feature selection processes are employed to reduce the number of features used to describe the system. This reduction can be accomplished by using only those descriptors that the user intuitively feels are the most relevant, by forming ratios or combinations of the original measurements, by applying various transforms to the data, or by using the results of the classification process to guide in selection of the most informative descriptors. If the selection process occurs before the classification step it is termed *prior*.

Alternatively, those processes that occur after the classification step are termed *posterior*. The latter are discussed further in Chapter 4.

Several types of prior methods have been detailed in the literature (46–50). Although each method differs in context, the concept is the same. Each method is based on obtaining the "optimal" set of descriptors. Optimality, however, is a relative quality that is dependent on the criteria that define it.

One method of finding optimal descriptors is to choose only those components that are most responsible for the clustering behavior of the class under study. The following method may be used to obtain these components. First we preprocess the data using the clustering transform **W**; that is, we form a new space **X*** in which the interclass distance for the points in one particular class is a minimum. As we saw before, this transform is simply

$$\mathbf{X}^* = \mathbf{WX} \tag{2.14}$$

where **W** is obtained from either equation 2.12 or 2.13. Associated with this space is a covariance matrix **C***, which can be formed from the original covariance matrix **C** through the transform

$$\mathbf{C}^* = \mathbf{WCW}' \tag{2.15}$$

where **W**′ is the transpose of the **W** matrix. Next we can decouple the covariance of the components in the **X*** space by finding **A**, the transform that maintains the minimum interclass distance and diagonalizes the covariance matrix. This results in a new space, **X****, in which the contribution of the various components to the clustering is evident.

We have created a new space in which we can evaluate the contribution of each of the features to the clustering of a particular class. Since features of larger variance contribute least to this clustering, those features would be ranked lower. Those features of smaller variance could be ranked higher. The n highest ranked features could be selected as those most relevant and others discarded.

Unfortunately, attempting to cluster the class before diagonalizing the covariance matrix would present problems in forming simple expressions for the transform matrices. However, if we first diagonalize the covariance matrix and then cluster the class of interest, the relationships become quite simple to form. If we proceed in this alternate order, it can be shown that the transform **A**, which diagonalizes the covariance matrix, is given by the matrix whose rows are the eigenvectors of the original covariance matrix **C**. Since **A** is orthonormal the original spatial relationships are not altered. The **W** transform can then be applied to maximize the clustering for the class of interest.

Preprocessing

A new space in which the clustering contribution of the features can be ranked according to their variance is developed by performing the following operations.

1. Calculate the covariance matrix **C** for the data using the relation

$$c_{ij} = \frac{1}{m-1} \sum_{l=1}^{m} (x_{il} - \bar{X}_i)(x_{jl} - \bar{X}_j) \qquad (2.16)$$

$$\bar{X}_k = \frac{1}{m} \sum_{l=1}^{m} x_{lk} \qquad (2.17)$$

2. Form the transform matrix A by finding the eigenvectors for C and using them as the rows of the A matrix.
3. Calculate the diagonalized covariance matrix using the relation

$$\mathbf{C^*} = \mathbf{ACA}^{-1} \qquad (2.18)$$

4. Using equations 2.9, 2.10, and 2.12 calculate **W**.
5. Calculate the covariance matrix for the double transformed space using

$$\mathbf{C^{**}} = \mathbf{WC^*W'} \qquad (2.19)$$

6. Transform the original data points into the new space using

$$\mathbf{X^{**}} = \mathbf{WAX} \qquad (2.20)$$

These operations yield a new data space in which the class of interest has the minimum interclass distance, and the data possess a diagonal covariance matrix. Those features having the lowest variance (diagonal entries of the covariance matrix) are considered the most relevant to the clustering of the data. The "optimal" subset is composed of the n members of lowest variance.

The most difficult part of these operations is calculation of the eigenvectors for the **A** transform. However, this can be accomplished for reasonably large data sets if efficient algorithms are used. The main disadvantage of this method is that the components in the **X**** space are not easily related to those of the original space because they are linear combinations of the original components. However, if it is not necessary to find the relationship in terms of the original components, then this method offers a viable approach to the feature reduction process if the original data set is truly representative of all possible data. Addition of new members to the data set is quite trivial, requiring only the multiplication of the new member's vector representation with the two transform matrices.

Another common definition of optimal is the formation of a reduced set of orthogonal basis vectors that reproduce the original distribution with the least amount of error. This type of transform is equivalent to finding the combination of rotations and projections necessary to decrease the total dimensionality while deviating from the original distribution the least. One method of obtaining such a reduced basis is the Karhunen–Loeve transform (14,51,52). If only a new basis is chosen for the data set no net reduction in the number of features is effected. However, the Karhunen–Loeve transform develops a basis by finding a reduced number of vectors that account for the major portion of the variance, as well as providing a distribution that deviates from the original with the least amount of error. In terms of providing orthogonal vectors with maximum variance retention and minimum distortion of the original distribution, the Karhunen–Loeve transform is optimal. This transform is quite easy to apply. A step by step procedure for the method is outlined below.

1. Compute the autocorrelation matrix, Q, from the patterns of the training set using the relation

$$Q = \frac{1}{T \cdot M_1} \sum_{j=1}^{M_1} \mathbf{X}_{1j} \mathbf{X}'_{1j} + \frac{1}{T \cdot M_2} \sum_{j=1}^{M_2} \mathbf{X}_{2j} \mathbf{X}'_{2j} + \cdots + \frac{1}{T \cdot M_k} \sum_{j=1}^{M_k} \mathbf{X}_{kj} \mathbf{X}'_{kj} \tag{2.21}$$

where M_k = the number of members in class k
T = the total number of classes
$\mathbf{X}_{kj}$ = the jth member of class k

For those who are unfamiliar with the notation XX', this is defined as follows. Given an n-dimensional vector $\mathbf{X}$ written

$$\mathbf{X} = \begin{bmatrix} x_1 \\ x_2 \\ x_3 \\ \vdots \\ x_n \end{bmatrix}$$

the transpose of which is

$$\mathbf{X}' = (x_1, x_2, x_3, \ldots, x_n)$$

then $\mathbf{XX}'$ is given by

$$\mathbf{XX}' = \begin{bmatrix} x_1 x_1 & x_1 x_2 & x_1 x_3 & \cdots & x_1 x_n \\ x_1 x_2 & x_2 x_2 & x_2 x_3 & \cdots & x_2 x_n \\ x_1 x_3 & x_2 x_3 & x_3 x_3 & \cdots & x_3 x_n \\ \vdots & \vdots & \vdots & & \vdots \\ x_1 x_n & x_2 x_n & x_3 x_n & \cdots & x_n x_n \end{bmatrix} \tag{2.22}$$

Preprocessing

Clearly, $\sum \mathbf{XX}'$ is the sum of all such matrices. There are as many of these as there are members in the class. The calculation of this sum need not require the actual generation and storage of as many matrices as data points. The sum can be found by combining the necessary operations.

2. Calculate the eigenvectors and eigenvalues for the $\mathbf{Q}$ matrix. Normalize the eigenvectors.
3. Choose the K eigenvectors that correspond to the K largest eigenvalues of the $\mathbf{Q}$ matrix. These will form the basis for the new space.
4. Form the transform matrix $\mathbf{A}$ by using the eigenvectors chosen in step 3 as the rows of the $\mathbf{A}$ matrix.
5. Compute the new data vectors $\mathbf{X}^*$ using the relation

$$\mathbf{X}^* = \mathbf{A}\mathbf{X} \tag{2.23}$$

The $\mathbf{X}^*$ vectors are now the K dimensional representation ($K < N$) that minimizes the approximation error. Thus the number of features needed to describe the data has been effectively reduced. Note that if $K = N$ there would be no net reduction in features and the new data points would simply represent the old points after a coordinate rotation.

Note two features of this process. First, the general form of $\mathbf{Q}$ is given by

$$\mathbf{Q} = \sum_{i=1}^{T} P(W_i) E\{\mathbf{X}_i \mathbf{X}_i'\} \tag{2.24}$$

Thus this transform can be used for cases in which the probabilities of observing class W_i are not all equal. The procedure outlined above is based on the assumption that all such probabilities are equal. Secondly, the Karhunen–Loeve expansion assumes that $E\{\mathbf{X}_i\} = 0$; that is, that all pattern classes possess zero means. While this is not always the case, the transform can still be applied. Obviously, data in which the means are widely disparate will not be transformed optimally; however, the results may still be of use.

Application of the Karhunen–Loeve transform generates descriptors that are linear combinations of those originally produced by the transducer. Unfortunately, contributions for descriptors that may be of little use are also added to the system. Therefore, to use this process properly, only those measurements that were originally the most useful should be transformed.

The Karhunen–Loeve transform is not dependent on the underlying probability distribution of the data. However, should this distribution be known or estimable, then selection techniques employing this information could be of use. One such method is entropy minimization. The procedure presented for this technique assumes that the data are normally distributed with equal covariance matrices. If this is the case or nearly the case, then the entropy minimization procedure presented below may be of use. A detailed derivation can be found in references 14 and 53.

1. Estimate the mean vector and the covariance matrix using

$$\mathbf{m} = \mathbf{m}_i = \frac{1}{N_i} \sum_{j=1}^{N_i} \mathbf{X}_{ij} \qquad (2.25)$$

and

$$\mathbf{C} = \mathbf{C}_i = \frac{1}{N_i} \sum_{j=1}^{N_i} \mathbf{X}_{ij} \mathbf{X}'_{ij} - \mathbf{m}_i \mathbf{m}'_i \qquad (2.26)$$

where N_i = the number of samples in class i
$\mathbf{X}_{ij}$ = the jth member of the ith class of points. Note that $\mathbf{X}$ has dimension N

The notations $\mathbf{XX'}$ and $\mathbf{mm'}$ are as given in equation 2.22. Since the covariance matrices for all classes are assumed to be equal, $\mathbf{C}_i$ and $\mathbf{m}_i$ need only be computed for one class.

2. Calculate the eigenvalues and eigenvectors for the $\mathbf{C}$ matrix. Normalize the eigenvectors. Since the covariance matrix is always symmetrical (as a result of the $\mathbf{XX'}$ and $\mathbf{mm'}$ calculation) it is always possible to find a set of real, orthogonal eigenvalues.
3. Construct the transform matrix, $\mathbf{A}$, by using as the rows of the $\mathbf{A}$ matrix the eigenvectors having the K smallest eigenvalues.
4. Compute the new data vectors, $\mathbf{X}^*$, using the relation

$$\mathbf{X}^* = \mathbf{AX} \qquad (2.27)$$

The $\mathbf{X}^*$ vectors are the K dimensional representation ($K < N$) that minimizes the entropy of the system. The entropy minimization tends to cluster the data together as the space is reduced. Of course the success that this method achieves is dependent on how closely the data obey the assumptions of normal distribution and equal covariance matrices.

Several other techniques can be used to selectively eliminate the least informative descriptors. Such measures as divergence (14), the U statistic (54), the F statistic (55), and Fisher ratio (56) provide information useful in deciding which descriptors are most useful. Some of these methods make distributional assumptions concerning the data. If these assumptions are in error the results of the statistical measurements may not prove reliable. One further drawback is the requirement that all combinations of descriptors must be tested to select the "best" set. This is a bit impractical for sets of over 20 descriptors, as such computations require a factorial number of iterations. These operations would result in further degradation of the algorithms performance, since "near optimal" methods must be devised to reduce the number of calculations necessary to make a good choice of descriptors.

Often methods that are far less sophisticated provide useful information. Measuring the predictive ability of single features often proves informative. Single feature predictive abilities can be calculated from the following algorithm.

1. Order the value for each descriptor from highest to lowest.
2. Starting with the lowest value note the number of members per class above that point and the number below that point.
3. Repeat step 2 moving up one member each time until all members have been used.
4. Report the highest percentage of predictive ability that was measured and the highest percentage correct per class.

Descriptors whose discriminating ability is 90% or greater may be of little utility in performing an analysis. Such may be the case if a descriptor is contained solely or largely in a single class. Such an unbalanced distribution may indicate that the data set does not properly represent the classes.

Similarly, descriptors that yield extremely low predictive abilities may either contain no useful information or indicate the presence of a multimodal distribution. Additional information is necessary to determine which is the case.

The information obtained from the single feature prediction can also be computed for all pairwise multiples of descriptors. Significantly better predictive abilities may indicate that such a combination is of greater utility than the individual descriptors. This has the net effect of reducing the dimensionality of the data by one for each combination.

Calculation of several single feature statistics can also aid in interpreting the utility of a single feature. Calculation of means, standard deviations, highest value, lowest value, and the total number of nonzero values is easily accomplished for each class. By making it possible to judge the information content of the data, these measurements aid in decisions as to whether the inclusion of a descriptor is warranted.

The correlation coefficient is another useful gauge. Highly correlated descriptors may contain essentially the same information. If several descriptors are highly correlated, then one of them may be chosen in hopes of providing a reduction in the number of features while retaining the same amount of information.

The utility of the above methods is dependent on the data set. If classification is the main concern, then transformations such as the Karhunen–Loeve transformation are useful. If the data conform to well-known distributions, then statistical tests will uncover the useful features. If not, then measurements such as single feature predictive abilities could provide information on which to base a decision to include or exclude a descriptor. Ultimately the results of classification determine whether the choice of preprocessing and feature selection was correct.

Multidimensional Scaling and Non-linear Mapping

Before he embarks on an extended analysis it is occasionally beneficial for the scientist to obtain a feeling for the structure of the data, that is, how the data is geometrically distributed in the hyperspace formed by its description vectors. Examples of structure are hyperspheroidal or hyperellipsoidal clusters and their degree of separability or overlap. An exact picture of data structure cannot be obtained by direct inspection, as it is not possible to preceive geometric representations of more than three dimensions. However, judicious selection of the manner in which the data are projected from the high dimensional space to the more easily visualized two- or three-dimensional space often results in a useful indication of structure.

Developing a feeling for the underlying structure of the data is often of great aid in deciding which type of classification procedure is warranted. An evident lack of underlying structure would indicate that the description being employed is providing little or no information concerning the data. There are two basic methods for developing lower dimensional representations, linear and nonlinear.

The simplest linear plots are variable by variable plots. These are the most trivial of the possible plots and obviously suffer from the fact that single variables generally are poor indicators of the overall performance of a set of variables. In addition, no provision exists for the exclusion of variables that contain a minimum amount of discriminatory information. Such plots, however, can give an extremely rudimentary concept of relations within the data set.

A simple improvement of the situation can be effected by use of rotation operators to provide real time two- or three-dimensional views of the data. This technique does not provide much of an advantage over simple variable by variable plotting unless there is an overriding reason to expect a certain rotation to be fruitful.

A more sophisticated approach in the selection of a rotation projection combination is to use a modification of the Karhunen–Loeve transform to provide a two- or three-dimensional representation that most closely approximates the distribution of the data. Using this procedure is analogous to finding the rotation that, when projected onto a lower dimensional space, yields a representation that maintains the maximum amount of variance and provides the minimum distortion of the original distribution. In this case

Preprocessing

the transform is applied to the entire data set rather than to one particular class. This transform is easily applied using the following steps:

1. Calculate the n means for each measurement using the relation

$$\bar{X}_k = \frac{1}{m} \sum_{l=1}^{m} x_{kl} \quad (2.28)$$

where m is the number of data points and n is the number of measurements for each point.

2. Calculate the covariance matrix **C**. Each element, c_{ij}, of the covariance matrix is obtained from

$$c_{ij} = \frac{1}{m-1} \sum_{l=1}^{m} (x_{il} - \bar{X}_i)(x_{jl} - \bar{X}_j) \quad (2.29)$$

3. Find the eigenvectors, u_k, and the eigenvalues, λ_k, for the covariance matrix.

4. Display the data using as the new coordinate axis the K (≤ 3) eigenvectors having the largest eigenvalues.

The amount of information contained in these vectors can be expressed as

$$P_v = \frac{100 * \sum_{i=1}^{K} \lambda_i}{\sum_{i=1}^{N} \lambda_i} \quad (2.30)$$

which represents the percent variance retained in this representation of the data. A small value for P_v indicates that only a small portion of the data set variability is contained in this representation and, therefore, the representation may not be accurate.

While this type of display finds an optimal rotation and projection, the nature of such projections often results in points that were originally separated in the hyperspace representation being overlayed in the projected representations. This can lead to problems in interpreting the structure of the data. Nonlinear methods attempt to minimize this problem by providing for a different method of dimensionality reduction.

Nonlinear methods of display are represented by the techniques of nonlinear mapping or multidimensional scaling. The origins of these methods stem from the work of people such as Kruskal (57,58), Sheppard (59), and Sammon (60). A large body of literature has developed describing these techniques and their use (61–65). The basic idea is to project the data into a lower dimensional space (usually 2 or 3) such that the maximum amount of similarity or dissimilarity between the data points is retained. The data structure is represented in the lower dimensional space by distributing the data such that there exists a minimum difference between the similarity of the lower and higher dimensional representations. Any measure of similarity may be used, however, the most common measure employed is that of distance. Generally Euclidian distance is used unless there is some overriding reason to employ other metrics. The objective is to perform a mapping from hyperspace to two or three space such that the distances, d_{ij}, in the original space deviate least from the distances in the lower space, d_{ij}^*. The error in such a projection is then simply the difference between the distances in the original space and those in the reduced space.

It is convenient to represent the deviations of the d_{ij}^* distances from the original d_{ij} distances by criterion functions that are invariant to ridged body motion of the configurations, as well to as dilations of the points. Three such sums of squared error functions are given below.

$$J_1 = \frac{1}{\sum_{i<j} d_{ij}^2} \sum_{i<j} (d_{ij}^* - d_{ij})^2 \tag{2.31}$$

$$J_2 = \sum_{i<j} \left(\frac{d_{ij}^* - d_{ij}}{d_{ij}}\right)^2 \tag{2.32}$$

$$J_3 = \frac{1}{\sum_{i<j} d_{ij}} \sum_{i<j} \frac{(d_{ij}^* - d_{ij})^2}{d_{ij}} \tag{2.33}$$

Note that each criterion function emphasizes a different facet of the error. J_1 emphasizes the largest errors, independent of the magnitude of the d_{ij}. J_2 emphasizes the largest fractional errors independent of the magnitude of $|d_{ij}^* - d_{ij}|$. J_3 is a compromise between the two, emphasizing the largest product of error and fractional error. J_3 was the criterion function selected by Sammon (60) as being most useful.

In practice the criterion functions can be minimized by any number of methods. If the Euclidian metric is used, the gradients for the criterion function become

$$\nabla_{y_k} J_1 = \frac{2}{\sum_{i<j} d_{ij}^2} \sum_{j \neq k} (d_{kj}^* - d_{kj}) \frac{y_k - y_j}{d_{kj}^*} \tag{2.34}$$

$$\nabla_{y_k} J_2 = 2 \sum_{j \neq k} \frac{d_{kj}^* - d_{kj}}{d_{kj}^2} \frac{y_k - y_j}{d_{kj}^*} \tag{2.35}$$

$$\nabla_{y_k} J_3 = \frac{2}{\sum_{i<j} d_{ij}} \sum_{j \neq k} \frac{d_{kj}^* - d_{kj}}{d_{kj}} \frac{y_k - y_j}{d_{kj}^*} \tag{2.36}$$

where y_i is the lower dimensional image of the ith sample point, that is, $d_{ij}^* = \|y_i - y_j\|$.

An optimal configuration can then be found by following standard gradient descent procedures. The initial configuration can be obtained by randomly spreading the points about in space, however, the Karhunen–Loeve transform can be used to obtain a more reasonable starting configuration. The K (≤ 3) vectors having the largest eigenvalues can be used as the axis for the initial distribution of the data. This has the added advantage of providing an estimation of the percent of variance retention in the initial configuration. This measure is calculated using equation 2.30.

Although extremely useful, these techniques are not without drawbacks. The interpoint distance matrix consists of $N(N-1)/2$ elements. Thus, depending on the type of computer available, it is limited to sets of approximately 250 to 450 data points. As with other techniques, the inclusion of measurements containing little or no information tends to obscure the relation contained in the data. Such noisy measurements may cause considerable overlap to be observed in the lower dimensional representation because the algorithm is forced to fit distance contributions from both meaningful and random measurements equally. Finally, the structure of very high dimensional spaces ($N > \sim 15$) are often simply too complex to be represented in two or three dimensions. In such cases the lower dimensional representations may exhibit severe overlapping when in reality the data are quite well separated.

Despite these problems, multidimensional scaling does offer a convenient method of picturing the structure of high dimensional data. Its use as a preprocessing tool allows the scientist to obtain some insight into which classification method may provide the best results. Another attractive feature of the scaling techniques is that their use is not restricted solely to preprocessing. In systems where $N \leq 3$ the most sophisticated pattern recognition system is the scientist himself. For cases in which the data's structure can be accurately represented in a lower space, scaling methods allow the scientist to decide on the class of an unknown(s) by visual inspection of its lower dimensional representation.

CLASSIFICATION

The procedures prior to classification attempt to encode information into numerical form. It is appropriate to pause here and reflect upon the consequences of the process used to make these transformations.

The basic concept in applications of pattern recognition is that making measurements on a system encodes information regarding the properties of that system. For example, measuring the lipophilic properties of a molecule may yield information concerning the compounds ability to interact with a biosystem. The vector representation of these measurements results in a distribution of points in space in which each point represents one element in the set. If the measurements are truly related to observable properties, then those elements that have the desired property tend to cluster in limited regions of this space, while those that do not, cluster in another. These regions may not be unique but may overlap. If the overlap is severe then no information is present.

The assumption that informative measurements cause clustering is central to the application of pattern recognition. The whole purpose of transduction, preprocessing, and prior feature selection is to develop descriptions of objects such that they cluster into classes that relate to the presence or absence of a property. Pattern recognition is in essence the elucidation of similarity within sets of data. Classification is the process responsible for developing the relations that define this similarity. There are numerous methods of developing such rules, however, most, if not all, present pharmacological applications have used nonparametric methods as the primary method of classifier development. To understand the basis for the nonparametric methods a short introduction to the parametric methods is necessary.

Parametric Classification Methods

Parametric methods are based on Bayesian statistics. These methods develop their classification rules directly from the probabilistic distribution of the data. These distributions arise as a result of the number and type of transducers used, as well as the types of preprocessing or prior feature selection methods employed. The goal of classification is to maximize the probability of correct classification by developing a function that defines the boundaries between the different classes. There are several means of approaching the problem of parametric decision function development. A simple example is given to demonstrate the basics of the process.

A straightforward classifier can be developed using the Bayes formula:

$$P(\mathbf{X})P(W_i/\mathbf{X}) = P(W_i)P(\mathbf{X}/W_i) \quad (2.37)$$

In this equation $\mathbf{X}$ is a vector whose components are derived from the various transducers. The numerical values of the components are responsible for the N space distribution of the data. The function $P(\mathbf{X})$ describes the probability of observing any given member of the data set, regardless of the class in which it belongs. $P(W_i)$ is the probability function that describes the distribution of the classes; that is, it is the odds for the occurrence of class W_i. $P(W_i/\mathbf{X})$ is the class conditional probability. It describes the probability that the class is W_i given that the point is $\mathbf{X}$. $P(\mathbf{X}/W_i)$ is the conditional probability for $\mathbf{X}$,

Classification

and can be considered the complement of $P(W_i/X)$. It describes the probability of picking **X** given that it is drawn from class W_i.

We can use the conditional probabilities to develop a discriminant function. A discriminant function, $f(\mathbf{X})$, is one that assigns each **X** to exactly one of the classes W_i. An optimal function is one that has the highest probability of being correct. The probability that **X** comes from class W_i is

$$P_i = \frac{P(\mathbf{X}/W_i)}{\sum_{k=1}^{M} P(\mathbf{X}/W_k)} \qquad (2.38)$$

This is equivalent to assuming that the *a priori* probabilities for each class are equal and, therefore, comes directly from the Bayes formula. Obviously, the largest value of $P(\mathbf{X}/W_i)$ is synonymous with the optimal decision function. This rule can be stated as X belongs to class W_i if

$$P(\mathbf{X}/W_i) > P(\mathbf{X}/W_j) \qquad j \neq i \qquad (2.39)$$

Alternatively stated,

$$\frac{P(\mathbf{X}/W_i)}{P(\mathbf{X}/W_j)} > 1 \qquad j \neq i \qquad (2.40)$$

In the case of both probabilities being equal, **X** can be assigned to either class W_i or class W_j.

A decision function can now be developed using equation 2.40. If we assume each class to be normally distributed with equal covariance matrices **C**, then the expansion for $P(\mathbf{X}/W_i)$ becomes

$$P(\mathbf{X}/W_i) = \frac{1}{(2\pi)^{n/2}|\mathbf{C}_i|^{1/2}} \exp\left[-\tfrac{1}{2}(\mathbf{X} - \mathbf{m}_i)' \mathbf{C}_i^{-1}(\mathbf{X} - \mathbf{m}_i)\right] \qquad (2.41)$$

where m_i is the mean vector. The ratios of the density function then become

$$\frac{P(\mathbf{X}/W_i)}{P(\mathbf{X}/W_j)} = \exp\left\{-\tfrac{1}{2}[(\mathbf{X} - \mathbf{m}_i)'\mathbf{C}^{-1}(\mathbf{X} - \mathbf{m}_i) - (\mathbf{X} - \mathbf{m}_j)'\mathbf{C}^{-1}(\mathbf{X} - \mathbf{m}_j)]\right\} \qquad (2.42)$$

Since **C** is symmetric, $\mathbf{X}'\mathbf{C}^{-1} = \mathbf{C}^{-1}\mathbf{X}$ and equation 2.42 reduces to

$$\frac{P(\mathbf{X}/W_i)}{P(\mathbf{X}/W_j)} = \exp\left[\mathbf{X}'\mathbf{C}^{-1}(\mathbf{m}_i - \mathbf{m}_j) - \tfrac{1}{2}(\mathbf{m}_i + \mathbf{m}_j)'\mathbf{C}^{-1}(\mathbf{m}_i - \mathbf{m}_j)\right] \qquad (2.43)$$

If we define

$$f_{i,j}(\mathbf{X}) = \ln \frac{P(\mathbf{X}/W_i)}{P(\mathbf{X}/W_j)}$$

we obtain the recognition function

$$f_{i,j}(\mathbf{X}) = \mathbf{X}'\mathbf{C}^{-1}(\mathbf{m}_i - \mathbf{m}_j) - \tfrac{1}{2}(\mathbf{m}_i + \mathbf{m}_j)'\mathbf{C}^{-1}(\mathbf{m}_i - \mathbf{m}_j) = 0 \qquad (2.44)$$

To classify **X** we compute $f_{ij}(\mathbf{X})$ for all i and j, $i \neq j$, assigning **X** to the class for which the greatest value of $f_{ij}(\mathbf{X})$ is obtained. This is the optimal decision function given the assumptions of equal *a priori* probabilities, equal covariance matrices, and normal distribution.

Note that $f_{ij}(\mathbf{X})$ can alternately be viewed as a hyperplane. In the two-class case, the rule for classification would be

$$f_{ij}(\mathbf{X}) > 0 \quad \text{for} \quad \mathbf{X} \in W_i$$

$$f_{ij}(\mathbf{X}) < 0 \quad \text{for} \quad \mathbf{X} \in W_j$$

It is clear that $f_{ij}(\mathbf{X}) = 0$ defines a decision boundary that separates class 1 from class 2. This function is called a decision function. Further details concerning development of decision functions from probabilistic arguments can be obtained from the references given previously.

It is interesting to note that the optimal function in this case is linear. This is not a surprising result if the distribution is viewed from a nonparametric perspective. Figure 2.5 shows a representation of the density functions for a two-class problem. This example shows a one-dimensional distribution in which the vertical axis represents the number of observations and the horizontal axis represents the values for the vector component. Note that the surface at position A has an error in classification that is proportional to the sum of the shaded and the cross hatched areas, whereas the surface at position B has an error that is proportional to the cross hatched area only. It is easy to see that the function describing B is the optimal decision function for this problem. No other surface has a lower probability

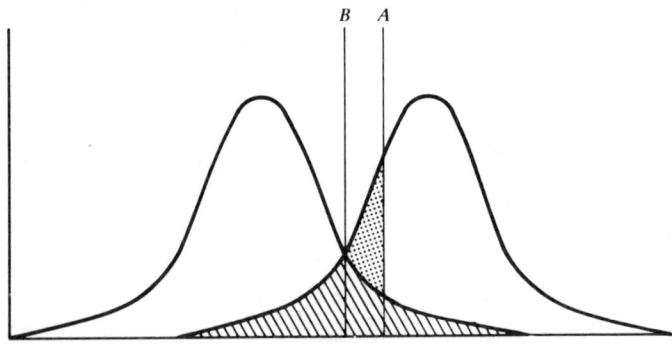

Figure 2.5 One-dimensional probability distributions showing optimal and nonoptimal Separating surfaces.

Classification

of misclassification. Indeed, equations 2.43 and 2.44 are seen to be the representation of this surface.

It is clear, then, that knowing the form of the distribution for $P(X/W_i)$ leads directly to the generation of a decision function. Generally parametric decision functions arise from the two different forms shown below:

$$f(X) = P(X/W_i)P(W_i) \qquad (2.45)$$

$$f(X) = P(W_i/X) \qquad (2.46)$$

Functions using these forms can be derived by use of elementary information theory in combination with the Bayes relation. The relation derived as an example is a special case of equation 2.45, which does not make the assumption that all classes are equally likely. This was one of the assumptions implicit in the derivation of equation 2.44.

Clearly, knowing the form of the distribution for $P(X/W_i)$, $P(W_i)$, or $P(W_i/X)$ will lead to an equation that is optimal in the sense of minimizing the probability of misclassification. Generally the forms of the distribution are not known, but are approximated. It is well known that many of these approximations yield decision surfaces that are linear, or at worst quadratic.

In effect, parametric methods are those that approximate the form of the distribution function to derive the form of the optimal decision function. If the approximation of the proper distribution function is in error, then, likewise, the decision function will perform suboptimally. The approximation of these distribution functions is the greatest barrier to the use of parametrically derived decision functions in many applications. Not only are assumptions concerning the form of the distribution often invalid, but the cost of their estimation in terms of computational time and effort is prohibitive.

As real data are often complex, and therefore assumptions concerning its distribution are difficult, another approach to the mechanism of decision function development must be taken. Note that if the form of the distribution function cannot be easily approximated, then perhaps the form of the decision function can be approximated. The problem then becomes one of properly placing the decision function so as to attain the closest approximation to the surface that would have been derived if the standard Bayesian approach had been used.

Classifiers that develop decision functions based on approximating the form of the decision surface are termed nonparametric classifiers. Generally these classifiers develop linear functions by finding the position at which the classification error of the classifier is a minimum. The accuracy of this method depends on how closely the data used to develop the classifier reflect the data's ultimate distribution. Chapter 4 deals with the capabilities and limitations of several types of nonparametric classifiers.

CLUSTERING CONCEPTS

One of the most intuitively appealing aspects of the classification problem is that of cluster definition. This approach naturally arises as a result of the geometric nature of the problem. This is especially clear for the example used earlier, where the problem was to define the boundaries of the interface between the cluster of normal and abnormal cells. Since the problem was low dimensional in nature, visually searching for clusters offered an intuitive solution to the problem. Clearly, such visual methods of cluster definition are limited to problems in which three parameters or less are involved. Additionally, visual definition often restricts the ability to use the classifier in a recognition process. Thus methods have been developed that provide for a more organized approach to the definition of clusters.

There are several algorithms available that divide the data set into a number of clusters. Most all of these depend on the use of various sorts of distance metrics to provide the measures of similarity used for the clustering process. The motivation for using such measures lies in the fact that distance is a natural tool for establishing similarity between vectors that we consider as points in Euclidean space. Distance measures, however, are only one of several ways in which clusters can be defined. Hartigan (66) has defined six different types of cluster algorithms as based on the mode in which a cluster is developed. These are given below.

SORTING   Sorting algorithms arrange the data in case by variable form. Each variable covers a limited range of distinct values. An important variable is chosen in some manner and the members of the data set are partitioned according to the values of this variable. Within each of the clusters of the partition, further partitioning is effected according to other important variables.

SWITCHING   Switching algorithms partition the data set into clusters by choosing a set of initial clusters and then switches an object from its membership in an initial cluster to another. This proceeds until some criterion (such as standard deviation per cluster) is reached. Switching algorithms are rapid to execute but are often restricted by the manner in which the initial clusters are defined.

JOINING   Joining algorithms operate by initially choosing a set number of clusters, each containing only one member of the data set. The closest pair of clusters are found and these are joined into a new, larger cluster. The joining is continued until all members have been assigned a cluster. The clusters reach optimal conditions, or the entire data set is contained in one cluster. For large data sets (>1000 members) this pairwise joining is not practical; thus approximations are needed to determine optimal conditions.

SPLITTING   Splitting algorithms are the exact opposite of joining algorithms. Here the entire data set is continually split into smaller clusters according to some splitting rule (minimal/maximal size, standard deviation, etc.). These algorithms generally have difficulty in deciding the exact form of the splitting functions.

ADDING   Adding algorithms sequentially add members of the data set into clusters (such as partitions or trees) that already exist. Its limitations are obvious.

SEARCHING   Searching algorithms are generally applied to systems in which mathematical considerations have ruled out many of the possible clusterings. These algorithms then attempt to form optimal clusters such that the cluster formation causes a minimum in an error function.

While many possible algorithms exist, no one algorithm is suited to every problem. Some algorithms, such as the ISODATA algorithm of Ball and Hall (67,68) provide for adding, searching, joining, and splitting. These types of algorithms do tend to have broader utility but are by no means globally applicable. Also, many of the cluster methods are heuristic in nature and

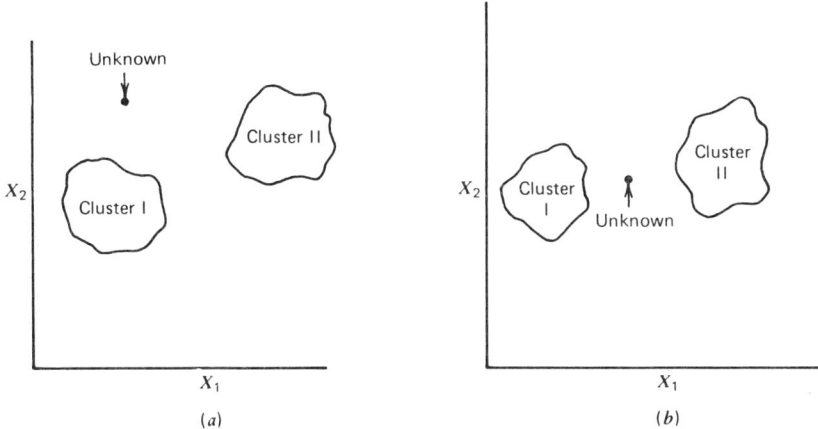

Figure 2.6 An unknown easily classified by a distance measure. (b) An unknown that is not easily classified by a distance measure.

thus their ultimate performance is dictated by the cleverness of the person developing them. A final drawback to the use of clustering is the problem of how to use the classification provided by the clusters to develop useful recognition processes. Figure 2.6 demonstrates this problem. Clearly, the unknown member in Figure 2.6 is easily placed in the cluster to which it is nearest. However, how does one intelligently place the unknown in Figure 2.6*b*? In cases such as this the distance from an unknown to the clusters themselves is of little utility in providing reliable recognition.

Although clustering has its drawbacks, it is useful in attempting to detect order in a system that initially appears to be largely disordered. Note also that these concepts do not require the points to be grouped into classes initially, thus clustering algorithms can be used to define classes for cases in which division of the data set into classes is not intuitively obvious. As we mention earlier, clustering algorithms that employ various distance metrics can be used to provide similarity indices, for developing definitions of useful features, or for transforming the data into forms more amenable to discriminant analysis.

REFERENCES

1. H. C. Andrews, *Introduction to Mathematical Techniques in Pattern Recognition*, Wiley, New York, 1972.
2. A. G. Arkadev and E. M. Braverman, *Learning in Pattern Classification Machines*, Nauka, Moscow, 1971.
3. B. G. Batchelor, *Practical Approach to Pattern Classification*, Plenum, New York, 1974.
4. P. W. Becker, *An Introduction to the Design of Pattern Recognition Devices*, Springer, New York, 1971.
5. C. H. Chen, *Statistical Pattern Recognition*, Hayden, New York, 1973.
6. R. O. Duda and P. E. Hart, *Pattern Classification and Scene Analysis*, Wiley, New York, 1973.
7. K. S. Fu, *Syntactic Methods in Pattern Recognition*, Academic, New York, 1974.
8. K. Fukanaga, *Introduction to Statistical Pattern Recognition*, Academic, New York, 1972.
9. W. Meisel, *Computer-Oriented Approaches to Pattern Recognition*, Academic, New York, 1972.
10. M. Minsky and S. Papert, *Perceptrons*, MIT Press, Cambridge, 1969.
11. N. J. Nilsson, *Learning Machines*, McGraw-Hill, New York, 1965.
12. E. A. Patrick, *Fundamentals of Pattern Recognition*, Prentice-Hall, Englewood Cliffs, New Jersey, 1972.
13. G. S. Sebestyen, *Decision Processes in Pattern Recognition*, Macmillan, New York, 1962.
14. J. T. Tou and R. C. Gonzalez, *Pattern Recognition Principles*, Addison-Wesley, New York, 1974.
15. L. Uhr, *Pattern Recognition, Learning and Thought*, Prentice-Hall, Englewood Cliffs, New Jersey, 1973.

References

16. J. R. Ullman, *Pattern Recognition Techniques*, Crane, Russak, and Co., New York 1973.
17. T. Y. Young and T. W. Calvert, *Classification, Estimation, and Pattern Recognition*, Elsevier, New York, 1973.
18. V. L. Tal'roze, V. V. Raznikov, and G. D. Tantsyrev, *Dok. Akad. Nauk SSSR*, **159**(1), 182 (1964).
19. P. C. Jurs, B. R. Kowalski, and T. L. Isenhour, Computerized Learning Machines Applied to Chemical Problems. Molecular Formula Determination from Low Resolution Mass Spectrometry, *Anal. Chem.*, **41**, 21 (1969).
20. P. C. Jurs and T. L. Isenhour, *Chemical Applications of Pattern Recognition*, Wiley-Interscience, New York, 1975.
21. B. R. Kowalski and C. F. Bender, Solving Chemical Problems with Pattern Recognition, *Naturwissenschaften*, **62**, 10 (1975).
22. B. R. Kowalski, Measurement Analysis by Pattern Recognition, *Anal. Chem.*, **47**, 1152A (1975).
23. P. C. Jurs, *Proceedings of the Workshop on Chemical Applications of Pattern Recognition*, Washington, D.C., May 1975.
24. T. F. Lam, C. L. Wilkins, T. R. Brunner, L. J. Soltzberg, and S. L. Kaberline, Large-Scale Mass Spectral Analysis by Simplex Pattern Recognition, *Anal. Chem.*, **48**, 1768 (1976).
25. H. Rotter and K. Varmuza, Computer-Aided Interpretation of Steroid Mass Spectra by Pattern Recognition Methods. Part 2. Influence of Mass Spectral Preprocessing on Classification by Distance Measurement to Centers of Gravity, *Anal. Chim. Acta*, **95**, 25 (1977).
26. S. R. Lowry, T. L. Isenhour, J. B. Justice, Jr., F. W. McLafferty, H. E. Dayringer, and R. Venkataraghavan, Comparison of Various K-Nearest Neighbor Voting Schemes with the Self-Training Interpretive and Retrieval System for Identifying Molecular Substructures from Mass Spectral Data, *Anal. Chem.*, **49**, 1720 (1977).
27. H. B. Woodruff, G. L. Ritter, S. R. Lowry, and T. L. Isenhour, Pattern Recognition Methods for the Classification of Binary Infrared Spectral Data, *Appl. Spectrosc.*, **30**, 213 (1976).
28. H. B. Woodruff and M. E. Munk, A Computerized Infrared Spectral Interpreter as a Tool in Structure Elucidation of Natural Products, *J. Org. Chem.*, **42**, 1761 (1977).
29. T. R. Brunner, C. L. Wilkins, R. C. Williams, and P. J. McCombie, Pattern Recognition Analysis of Carbon-13 Free Induction Decay Data, *Anal. Chem.*, **47**, 662 (1975).
30. T. R. Brunner, C. L. Wilkins, T. F. Lam, L. J. Soltzberg, and S. L. Kaberline, Simplex Pattern Recognition Applied to Carbon-13 Nuclear Magnetic Resonance Spectrometry, *Anal. Chem.*, **48**, 1146 (1976).
31. H. B. Woodruff, C. R. Snelling, Jr., C. A. Shelley, and M. E. Munk, Computer-Assisted Interpretation of Carbon-13 Nuclear Magnetic Resonance Spectra Applied to Structure Elucidation of Natural Products, *Anal. Chem.*, **49**, 2075 (1977).
32. M. Sjostrom and U. Edlund, Analysis of 13-C NMR Data by Means of Pattern Recognition Methodology, *J. Magn. Resonance*, **25**, 285 (1977).
33. Q. V. Thomas and S. P. Perone, Application of Pattern Recognition Techniques to the Interpretation of Severely Overlapped Voltammetric Data: Theoretical Studies, *Anal. Chem.*, **49**, 1369 (1977).
34. Q. V. Thomas, R. A. DePalma, and S. P. Perone, Application of Pattern Recognition Techniques to the Interpretation of Severely Overlapped Voltammetric Data: Studies with Experimental Data, *Anal. Chem.*, **49**, 1376 (1977).
35. J. R. McGill and B. R. Kowalski, Recognizing Patterns in Trace Elements, *Appl. Spectrosc.*, **31**, 87 (1977).

36. P. K. Hopke, The Application of Multivariate Analysis for Interpretation of the Chemical and Physical Analysis of Lake Sediments, *J. Environ. Sci. Health*, **A11(6)**, 367 (1976).
37. P. D. Gaarenstroom, S. P. Perone, and J. L. Moyers, Application of Pattern Recognition and Factor Analysis for Characterization of Atmospheric Particulate Composition in Southwest Desert Atmosphere, *Environ. Sci. Technol.*, **11**, 795 (1977).
38. P. L. Briggs and F. Press, Pattern Recognition Applied to Uranium Prospecting, *Nature*, **268**, 125 (1977).
39. J. S. Mattson, C. S. Mattson, M. J. Spencer, and F. W. Spencer, Classification of Petroleum Pollutants by Linear Discriminant Function Analysis of Infrared Spectral Patterns, *Anal. Chem.*, **49**, 500 (1977).
40. D. L. Massart and H. De Clerq, Application of Numerical Taxonomy Techniques to the Choice of Optimal Sets of Solvents in Thin Layer Chromatography, *Anal. Chem.*, **46**, 1988 (1974).
41. A. Eskes, F. Dupuis, A. Dijkstra, H. DeClercq, and D. L. Massart, Application of Information Theory and Numerical Taxonomy to the Selection of Gas-Liquid Chromatography Stationary Phases, *Anal. Chem.*, **47**, 2168 (1975).
42. J. K. Haken, M. S. Wainwright, and N. D. Phuong, The Nearest Neighbour Technique as a Means of Indicating Stationary Phase Selectivity, *J. Chromatogr.*, **117**, 23 (1976).
43. S. R. Lowry, G. L. Ritter, H. B. Woodruff, and T. L. Isenhour, Selecting Liquid Phases for Multiple Column Gas Chromatography from Their Eigenvector Projections, *J. Chromatogr. Sci.*, **14**, 126 (1976).
44. B. G. M. Vandeginste, Pattern Recognition as a Procedure for Selecting Analytical Methods for Solving Analytical Problems. A Preliminary Investigation, *Anal. Lett.*, **10(9)**, 661 (1977).
45. R. L. Shriner, R. C. Fuson, and D. Y. Curtin, *The Systematic Identification of Organic Compounds*, Wiley, New York, 1967.
46. Ho Y-C and R. L. Kashyap, An Algorithm for Linear Inequalities and Its Application, *IEEE Trans. Elect. Comp.*, **EC-14**, 683 (1965).
47. Julius T. Tou, *Computer and Information Sciences III. Proceedings of the Second Symposium on Computer and Information Science*, Battelle Memorial Institute, August 22–24, 1966, Academic, New York, 1967.
48. Second International Joint Conference on Pattern Recognition, August 13–15, 1974, Copenhagen, Denmark, IEEE Cat. No. 74CHO885-4C.
49. Third International Joint Conference on Pattern Recognition, November 8–11, 1976, Coronado, California, IEEE Cat. No. 76CH1140-3C.
50. M. H. Von Emden, *An Analysis of Complexity*, Mathematical Centre Tracts, Mathematisch Centrum, Amsterdam, 1971.
51. A. N. Mucciardi and E. E. Gose, A Comparison of Seven Techniques for Choosing Subsets of Pattern Recognition Properties, *IEEE Trans. Comp.*, **C-20**, 1023 (1971).
52. H. C. Andrews, Multidimensional Rotations in Feature Selection, *IEEE Trans. Comp.*, **C-20**, 1045 (1971).
53. J. T. Tou and R. P. Heydorn, *Some Approaches to Optimum Feature Extraction*, Computer and Information Science II, Academic, New York, 1967, p. 57.
54. G. P. McCabe, Computations for Variable Selection in Discriminant Analysis, *Technometrics*, **17**, 103 (1975).
55. P. R. Bevington, *Data Reduction and Error Analysis for the Physical Sciences*, McGraw-Hill, New York, 1969.

References

56. R. A. Fisher, The Use of Multiple Measurements in Taxonomic Problems, *Ann. Eugen.*, **7**, 178 (1936).
57. J. B. Kruskal, Multidimensional Scaling by Optimizing Goodness of Fit to a Numeric Hypothesis, *Psychometrika*, **29**, 1 (1964).
58. J. B. Kruskal, Nonmetric Multidimensional Scaling: A Numerical Method, *Psychometrika*, **29**, 115 (1964).
59. R. N. Sheppard, The Analysis of Proximities: Multidimensional Scaling with an Unknown Distance Function, *Psychometrika*, **27**, 125 (1962).
60. J. W. Sammon, Jr., A Nonlinear Mapping for Data Structure Analysis, *IEEE Trans. Comp.*, **C-18**, 401 (1969).
61. P. E. Green and E. J. Carmone, *Multidimensional Scaling and Related Techniques in Market Analysis*, Allyn and Bacon, Boston, Massachusetts, 1970.
62. P. E. Green and V. R. Rao, *Applied Multidimensional Scaling—A Comparison of Approaches and Algorithms*, Holt Rinehart and Wilson, New York, 1972.
63. W. S. Torgenson, *Theory and Methods of Scaling*, Wiley, New York, 1958.
64. C. H. Coombs, *A Theory of Data*, Wiley, New York, 1964.
65. G. Young and A. S. Houscholder, Discussion of a Set of Points in Terms of Their Mutual Distances, *Psychometrika*, **3**, 19 (1938).
66. J. A. Hartigan, *Clustering Algorithms*, Wiley, New York, 1975.
67. G. H. Ball and J. P. Hall, ISODATA: A Novel Method of Data Analysis and Pattern Classification, NTIS Report AD699616. 1965
68. G. H. Ball and J. P. Hall, ISODATA, An Iterative Method of Multivariate Analysis and Pattern Classification, *Proceedings of the IFIPS Congress*. 1965

CHAPTER 3

Chemical Structure Information Handling: Molecular Descriptor Development

Before any structure–activity analysis can be performed, each member of the data set must be described in some manner. For Hansch analysis, hydrophobic, electronic, and steric parameters are used to describe each molecule. However, these descriptors are difficult, if not impossible, to obtain for a structurally diverse data set. Consequently, a large amount of effort has gone into searching for parameters that can produce acceptable results but can be readily obtained for any chemical structure. In this chapter, we describe various techniques for generating molecular descriptors from chemical structures and discuss some techniques for handling chemical structures with a computer.

BASIC CONCEPTS OF MOLECULAR STRUCTURE CODING

Efforts to keep abreast of the explosion of publications in chemistry in the past decade have centered on the development of computer methods for the handling of chemical structure information. Since the cornerstone of chemistry is the universal coding system of two-dimensional structural diagrams drawn on blackboards and paper, computer systems have had to be designed to deal with this type of structural representation. Since such a graphical representation is incompatible with the internal necessities of computers, new coding systems have been devised to eliminate this incompatibility.

Molecular structure representation methods can be broken down into a simple hierarchy as shown in Figure 3.1. Ambiguous representations store enough information about the chemical structure to be useful in certain situations, but not enough to allow the total reconstruction of the original molecule. A prime example of this type of representation is the molecular

Basic Concepts of Molecular Structure Coding

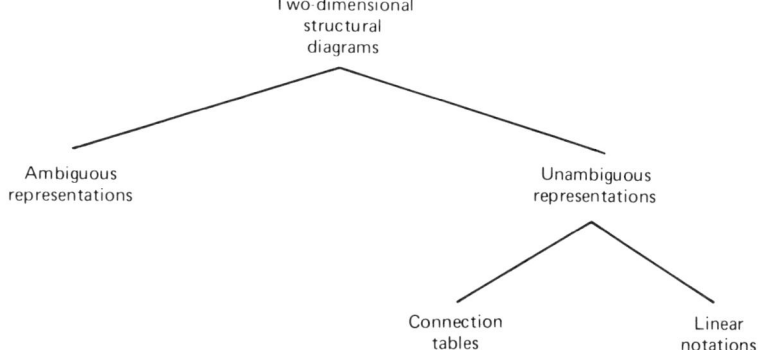

Figure 3.1 A hierarchy of molecular structure representations.

formula, in which information about the composition and relative weight of a molecule is maintained but the structural information is lost. Thus C_4H_8 could represent cyclobutane, 1-butene, or even isobutene. Another example of an ambiguous representation is fragment encoding where only portions of the structures of special interest (e.g., functional groups or ring systems) are represented.

Unambiguous representations provide a complete representation of the topology of the molecule, and they can be used to reconstruct the original molecule. Topological representations include two main approaches—linear notations and connection tables. Naturally, these two approaches have disadvantages that make neither one of them the best choice for all situations.

Linear notation methods represent chemical structures with strings of symbols. The term linear is used because the string of symbols can be prepared with a typewriter, linotype, or keypunch with the symbols occupying only one line. Two examples of linear notations that refer to the same compound, cyclohexanol, are L6TJV and A6EQ. For such linear notations to have meaning, they must be formed using an established set of rules, which are usually arbitrary.

Properties of linear notations of interest are uniqueness and specificity. Uniqueness refers to the ability of a notation to provide precise enough rules so that only one notation can be derived for a given structure. Specificity refers to the property that a notation yields only one structure upon interpretation. In the development of a particular notation system, tradeoffs must be made among uniqueness, specificity, and brevity, as well as between compactness of the notation and the complexity of the rules. A number of linear notation systems have been implemented, including the IUPAC, Wiswesser, Hayward, Skolnik, and GREMAS systems. To illustrate some features of a specific linear notation method, we use the Wiswesser line notation.

WISWESSER LINE NOTATION

The most often used linear notation method was originated in 1953 by W. J. Wiswesser (1). The most authoritative account of Wiswesser line notation (commonly referred to as WLN) has been presented by Smith (2). A Wiswesser line notation is constructed by applying a set of rules to the structure to generate a string of symbols, each corresponding to a structural fragment of the molecule. The sequence of symbols is predetermined by the rules. The notation generated is compact, since most bonds are cited implicitly, and large structural fragments, such as rings, are often coded with one or a few symbols. The 40 symbols used in WLN are all standard symbols available on typewriters and card punches: ƀ & -/0123...9ABCD...XYZ, where ƀ stands for a blank space. Emphasis on symbol choice is put on functional groups in WLN. Thus oxygen can be coded in one of four ways:

O   Oxygen connected to atoms other than H, for example, esters and ethers

Q   Hydroxyl group, —OH

V   Carbonyl group, $\diagdown$C=O$\diagup$

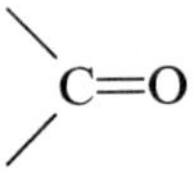

W   Dioxo group, such as —NO_2 or —SO_2—

Nitrogen can also be coded in one of four ways:

Z   —NH_2

M   —NH—

N 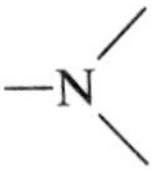   —N$\diagup\diagdown$

K   —N—+ (with vertical bonds above and below)

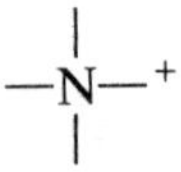

Carbons are coded as follows:

C   Unbranched carbon, doubly or triply bonded to at least one other element, for example, nitriles

Numeral   Straight chain alkyl group of indicated length

Y   Branched carbon with three connections

X   Branched carbon with four connections

Wiswesser Line Notation

Any halogen is coded as J while E, F, G, and I stand for bromine, fluorine, chlorine, and iodine, respectively. The symbol U denotes a double bond and UU denotes an acetylenic bond. The hydrogen atom is cited as H when it is not implied as part of another symbol or to finish the notation for a linear, unbranched hydrocarbon.

The derivation of the WLN for an acyclic structure is accomplished through the following sequence:

1. Find the linear sequence of notation symbols that contains the largest number of branching symbols or the largest number of symbols.
2. Write the sequence left to right, starting with the end of the string with the symbol that appears last in the symbol string listed previously.
3. Decide which of several symbols will appear in the sequence at branch points by a set of precedence rules.

An ampersand separates the completion of one branch and the start of a second one unless the first branch has a terminal group on its end. Examples of the WLN of five acyclic structures are shown in Table 3.1.

Table 3.1 Wiswesser Line Notation Examples

A.	$CH_3-CH_2-O-CH_2-CH_2-CH_3$	3O2
B.	$H_2N-CH_2-CH_2-\overset{\overset{O}{\|\|}}{C}-OH$	Z2VQ
C.	$CH_3-CH=N-NH-\overset{\overset{O}{\|\|}}{C}-NH_2$	ZVMNU2
D.	$HO-CH_2-\underset{\underset{NH_2}{\|}}{CH}-CH_2-CH_2-CH\begin{smallmatrix}\nearrow CH_2-NH_2\\ \searrow CH_2-OH\end{smallmatrix}$	Z1Y1Q2YZ1Q
E.	$CH_3-CH_2-CH_2-CH_2-N\begin{smallmatrix}\nearrow CH_2-CH_3\\ \searrow CH_2-CH_2-CH_3\end{smallmatrix}$	4N3 2

Notes. A: 3 is later than 2 in the precedence list. B: Z is later than Q in the precedence list. D: The longest chain containing the two Y branch points is found. The end with Z starts the linear notation. The 1Q branch on the first Y is cited next because it has no branch symbol. No ampersand is needed because Q is a terminal group symbol.

Cyclic structures have their own rules beyond those for branched acyclic structures. The benzene ring symbol is R and it is subordinated to all other notation symbols. Carbocyclic ring notations begin with L, and heterocyclic ring structure notations begin with T. Either L or T is followed by a numeral showing the number of atoms in the ring. The set of WLN rules for rings, fused rings, and bicyclic and tricyclic rings keeps growing from this point and it is far too large to cite here. However, further discussions of WLN with examples can be found in Davis and Rush (3) and Lynch et al. (4), with the complete WLN system being explained in detail by Smith (2).

Other forms of linear notation are similar to those of Wiswesser except the symbols and rules of use are different. As can be seen from this short discussion, linear notations are advantageous in that no special equipment is necessary for computer input; a linear string of characters is obtained for all structures, and, by following all the rules, a unique structural representation is obtained. The largest drawback for these types of techniques is the necessity of learning a complex set of rules in order to code and decode the structure. Although computer programs have been written to handle this coding, the algorithms are large and complex because of the number of rules involved in the notation scheme.

CONNECTION TABLES

Connection tables are explicit representations of molecular structures in which atom identities, bond types, and interconnections between atoms are cited individually. However, unlike linear notation schemes, the rules for forming connection tables are simple and are easily applied for the coding of any chemical structure.

Physically, a connection table is nothing more than a square matrix where the atom identities are stored in the main diagonal while the bonding information is stored in the off-diagonal elements. The value assigned to each element in the connection table is determined by the index number given to each atom in the structure and the numeric codes chosen to represent the different atom and bond types. To further explain the conventions of connection tables, an example is given.

In Table 3.2 the structural diagram for caffeine is shown along with its connection table. The numeric codes used in the connection table for the different atom and bond types are also given. The sequence of atom numbering in the structure is arbitrary and merely indicates the atom's location in the connection table. Hydrogen atoms are not explicitly listed, but they are assumed to fill all otherwise vacant valencies. The main diagonal of the connection table matrix contains the numeric code for each atom (i.e., $a_{1,1} = 1$

Table 3.2 An Example of the Connection Table Format

Atom No.	1	2	3	4	5	6	7	8	9	10	11	12	13	14
1	1	1				1					2			
2	1	1	2				1							
3		2	1	1				1						
4			1	3	1									1
5				1	1	1							2	
6	1				1	3						1		
7		1					3	1		1				
8							1	1	2					
9			1					2	3					
10							1			1				
11	2										2			
12						1						1		
13					2								2	
14				1										1

Atom Type	Numeric Codes	Bond Type	Numeric Codes
C	1	Single	1
O	2	Double	2
N	3	Triple	3
S	4	Aromatic	4
F	5	Delocalized	5
Cl	6	Ionic	6
Br	7		
I	8		
P	9		

because atom number 1 is a carbon). The off-diagonal elements of the connection table contain the bonding information for the structure. If the matrix element $a_{i,j}$ is greater than zero, a bond exists between atoms i and j, with the numerical value of the entry being equal to the bond type. A value equal to zero indicates the absence of a bond between atoms i and j.

The numeric codes employed in this illustration were chosen in an arbitrary fashion and do not represent a universal standard. However, this does not present a problem since any existing connection table can be quickly and easily transformed from one coding system into another by a simple computer program. This is not true for linear notation methods.

Methods for renumbering connection tables in an effort to achieve a unique "name" for compounds have been developed and reported, for example, see the methods of Gluck (5) and Morgan (6). The algorithm develops a canonical connection table regardless of the initial numbering sequence used for initial generation of the connection table. The resulting single, invariant, canonical connection table is then the name for the compound. The ability to generate such a unique name for a compound is desirable for updating and maintaining large files of structures. Given a unique way of naming compounds, it is simple to check a file for duplicates. While the algorithms used to generate canonical names do indeed handle a large majority of the structures encountered, it can be shown that they are not always practical. At this point, two structures must be compared one atom at a time to see if they are identical. The use of such algorithms in production environments, such as that of the Chemical Abstract Service, is economical in that it saves a great deal of searching and editing time.

Connection table representations of molecular structures can be stored in any of a large number of formats. As is pointed out above, the actual matrix formulation of the connection table is symmetric. Therefore, only the main diagonal and the upper triangle need to be stored for subsequent use. By renumbering the structure, and then by letting placement in lists convey information, significant reductions in the length of names can be achieved. For storage of connection tables in computer files, compression of the records can be done with chemical structure information just as with any binary data string.

STRUCTURE ENCODING

The transformation of the structural diagram into a computer compatable format is usually the rate determining step in any chemical information handling system regardless of the method chosen for structural representation. For the occasional handling of a few structures, the manual encoding and entering of the data into the computer is quite sufficient. However, for the handling of large data sets on a regular basis, manual encoding is too time consuming and plagued with errors. Therefore, it is preferable to use computer controlled graphical input that allows the chemist to sketch the molecular structure and obtain immediate visual results for checking. Since sketching is also the natural language of chemists when dealing with molecular structures, this operation is natural.

A number of such input routines have been implemented. Devices such as RAND tablets, light-pens coupled with cathode ray tubes, and a computer controlled television camera have all been used for structure encoding. One specific example is described here. This routine, UDRAW, was written for the input of molecular structures via a Tektronix storage cathode ray tube (CRT) computer graphics terminal. Like other graphical input methods, the only action required to enter a structure is the drawing of the structure on the screen; the computer representation of the structure is automatically created by the algorithm.

ROUTINE UDRAW

UDRAW (7) was developed using a Tektronix 4010-1 computer controlled display terminal and the Tektronix "PLOT-10" standard graphics software package. This display terminal is equipped with an input cursor that allows the operator to indicate any point on the display screen. This cursor permits the user to interact with the program and sketch chemical structural diagrams.

The display terminal is controlled by a 16-bit word MODCOMP II digital computer having 96K bytes of memory. UDRAW is written in standard FORTRAN and is independent of machine word size. It requires about 8.4K bytes of memory in its present form with the required PLOT-10 subroutines taking an additional 4K bytes of memory.

A general flow diagram for routine UDRAW is given in Figure 3.2. The first section initializes storage arrays and displays the directive menu. Once the first molecular diagram has been entered, the initialization section is bypassed and the previous entry is redrawn. Thus the operator has the choice of either modifying an existing structure or initializing the arrays and entering a new molecule. This feature allows for fast encoding of data sets containing similar molecules.

The procedure for encoding a structure such as acrylic acid is simple and straightforward. As the routine enters the connection table section of UDRAW, the cursor appears. The operator then moves the cursor by means of thumb-wheels to the desired position of the first atom. The space bar is then depressed to indicate a carbon atom and this is followed by a RETURN. The cursor reappears ready to accept the location of the second atom. After

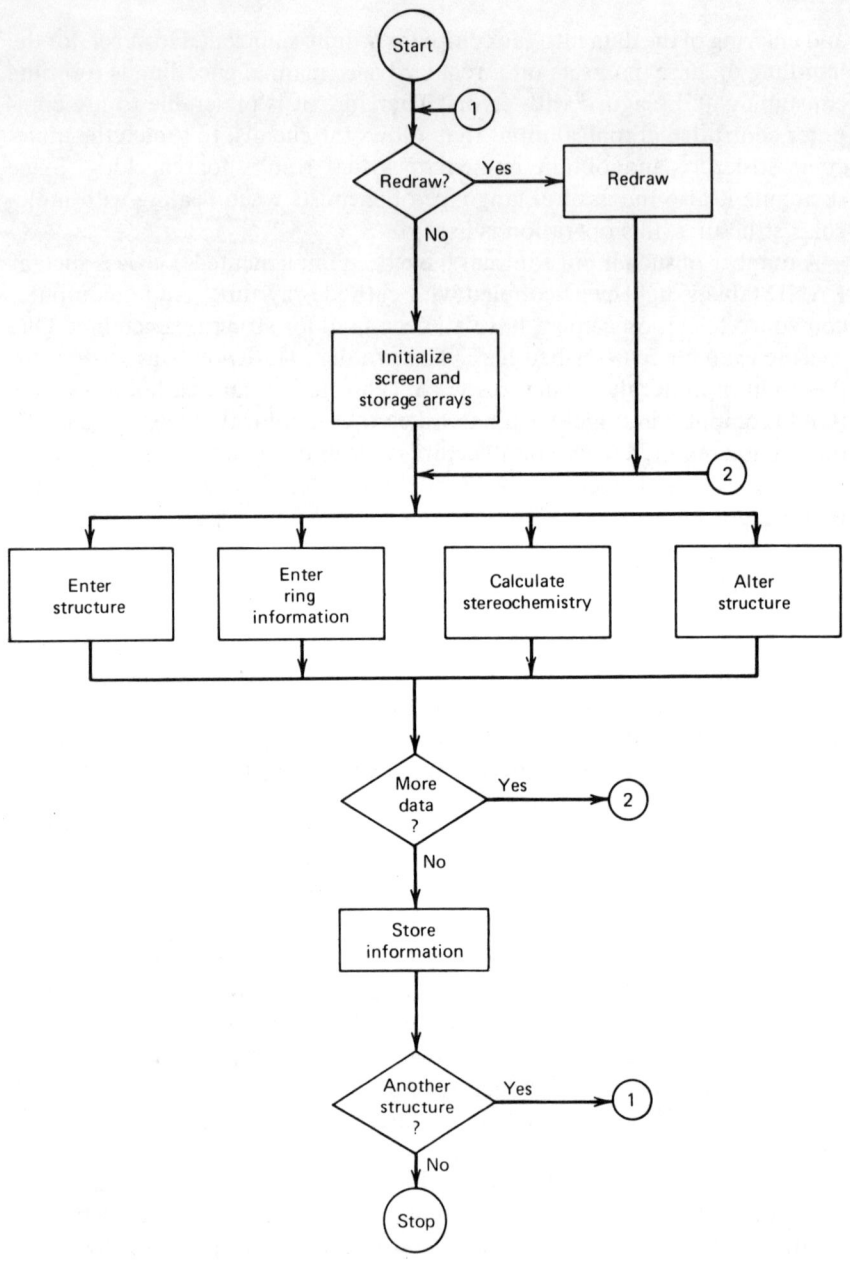

Figure 3.2 Flow chart of routine UDRAW.

moving the cursor to the desired location of the second atom, the user enters "Space"-RETURN to indicate another carbon atom. This is followed by depression of "2"-RETURN, which generates a carbon to carbon double bond. Since the atom last entered is a member of the next bond, the cursor is left in the previous position and "Space"-RETURN indicates this atom as the first atom to which the second bond will be connected. The cursor is then moved to the desired position of the third carbon atom of the molecule; "Space"-RETURN followed by "1"-RETURN generates a single bond between carbons 2 and 3. The double bond to the oxygen is generated by first pressing "Space"-RETURN while the cursor is on atom 3. Then, after the cursor is moved, "O"-RETURN is depressed to indicate an oxygen atom and "2"-RETURN designates a double bond. Since the final atom is also connected to atom 3, the cursor is first positioned over atom 3 and "Space"-RETURN is entered. This indicates the atom as the first of the final pair. The cursor is then positioned for the final atom and "O"-RETURN and "1"-RETURN generates the carbon to oxygen single bond. Since the structure is completed, a "D"-RETURN is entered, which causes the routine to branch to the directive input section of UDRAW. In this case the "FINISH" command is then given.

The display on the screen at the end of the input procedure would appear as follows if the molecule was entered from left to right:

$$C=C-C\underset{O}{\overset{O}{\diagup\!\!\!\!\diagdown}}$$

As a molecular structure is entered, a chemical connection table containing the identity of each atom, the connectivity of each atom, and the bond type of each connection is generated. Two-dimensional screen coordinates for each atom are also stored and are converted into angstrom coordinates in the final section of UDRAW. Atom and bond types that are recognized by UDRAW are given in Table 3.2. The keyboard character used to enter the atom is that shown for "atom type" with one exception; carbon is entered as a blank space. The bond type numeric codes are used for designating the bond type. Modification of this list requires only the changing of a few parameter statements. Thus the program is adaptable to a wide variety of molecular structures.

In addition to the connection table generation section, a ring information section and double bond stereochemistry section have been included. The information from these program modules may not be necessary for chemical data retrieval systems, but it is used in molecular modeling and substructure searching programs.

As the operator points out ring atoms with the cursor, the routine enters them into ring information storage arrays. The atom number of each ring atom and the size of the ring of which each ring atom is a member are stored for later use. If desired, a ring finding algorithm can be used instead of this manual procedure.

The section that calculates the stereochemistry about double bonds is completely operator independent once the structure has been entered. It consists of a search of all bonds to locate double bonds, a determination to make sure that the double bond is not a terminal bond, and finally a calculation to determine the stereochemistry about interior double bonds. To calculate the stereochemistry, the screen coordinates of the atoms are used. Thus the operator must draw the correct representation of the molecule on the screen. The calculation involved is the taking of the dot product of the single bonds attached to each atom forming the double bond. If 2-pentene was entered as

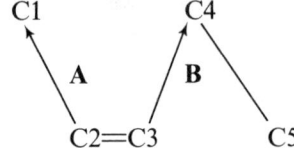

the bond going from C2 to C1 could be represented as vector **A**, and the bond going from C3 to C4 could be represented as vector **B**. The definition of the dot product of two vectors is $\mathbf{A} \cdot \mathbf{B} = |\mathbf{A}| |\mathbf{B}| \cos \theta$, where θ is the angle formed by the two vectors. Since $\cos \theta$ is negative for θ greater than 90°, the sign of the scalar quantity, $\mathbf{A} \cdot \mathbf{B}/|\mathbf{A}| |\mathbf{B}|$, indicates whether the bond is cis or trans. In this case, since $\cos \theta$ is greater than zero, the bond is cis.

Once a molecular structure has been entered, changes can be made using the alteration section of UDRAW. The operator can change any atom identity or any bond type. If desired, an atom can be entirely deleted from the structure. If the ring information is entered incorrectly, the errors are easily rectified. Also, double bonds drawn in one stereochemical configuration can be changed to the other type by simply indicating the bond to be changed. To verify that the alterations have been entered correctly, the entire molecular structure may be redrawn at any time.

Upon the input of the "FINISH" directive, the routine first checks flag variables that indicate whether the ring and stereochemistry sections have been entered by the user. If the flags indicate a negative answer, the section is entered and the necessary calculations are performed. After this, a set of coordinates is calculated for each atom using the screen coordinates.

Throughout the routine, checks are made to assure that acceptable information is being entered. Whenever an error is encountered, the display terminal's bell is rung and the routine branches back to the point where the error was made. Also, as the data is accepted, information is echoed back to the terminal in the form of symbols, bonds, or numbers indicating ring information. In this way constant visual feedback is presented to the user and incorrect data are not accepted.

Routine UDRAW is employed as the structural input routine in the structure–activity system described in a subsequent chapter of this book. It is used to input molecular structures into large data files, to input structures for molecular modeling, and to input substructures to be used in substructure searching. The routine exists in a stand-alone version and is available from the Quantum Chemistry Program Exchange (8).

MOLECULAR STRUCTURE DESCRIPTORS: TOPOLOGICAL

The development of molecular structure descriptors is the most important part of any structure–activity investigation because the descriptors must contain enough information to permit the correct classification of the compounds under study. In some instances enough prior knowledge exists in the form of a model to dictate which parameters are important for distinguishing objects belonging to different classes. However, in other applications descriptors are chosen on the basis of published experimental results.

Molecular descriptors can either be calculated using the structural information contained in the connection tables or obtained experimentally. The first procedure is the most attractive since it is the fastest and allows any chemical structure, either hypothetical or real, to be described.

The descriptors discussed in this section can be categorized as being topological in nature. Topological descriptors include fragment descriptors, which code the atom and bond types; substructure descriptors, which code the presence or absence of explicitly defined substructures; environment descriptors, which code the immediate surroundings of a substructure; and connectivity descriptors, which are indices of the extent of branching in the molecular structure. Geometrical descriptors, which are discussed in a subsequent section, are derived from strain minimized three-dimensional structures and code the general shape and size of the molecule. Most of these descriptors can be generated from linear notations of molecules, as well as from connection tables. However, the use of linear notations greatly increases the level of complexity for any given descriptor generating algorithm. Therefore, the following discussion is limited to molecules represented as connection tables.

Fragment Descriptors

The total number of atoms and bonds in a chemical structure provides some information about the composition of the compound. However, further information about the chemical nature can be obtained by fragmenting the structure into its basic atom and bond components.

The total number of atoms in a chemical structure can be subdivided into smaller groups by counting the number of times an element occurs in the structure. Since there is no advantage in including atom types not found in the data set under investigation, the number of atom fragment descriptors generated is data set dependent. Atom descriptors that can be used in studies of organic compounds are given in Table 3.3. The values for these descriptors are easily obtained from the main diagonal of the connection table. Because of the numeric codes employed for the atom types in the connection table, the sorting and counting of the atom descriptors is a trivial task for a digital computer.

An analogous sorting and counting procedure can be carried out on a chemical structure for the various bond types. However, the total number of new descriptors possible is smaller because of the limited number of different bond types. The six bond fragment descriptors given in Table 3.4 can describe most molecules. Since the number of exceptions found in any given data set is small, there is no advantage to the inclusion of other bond fragments. As a matter of fact, most data sets contain so few ionic and delocalized bonds that their information value may be too small to warrant their use in any study. Nevertheless, they are included for completeness.

In generating the bond fragments, a distinction is not made between identical bond types formed by different atom types, that is, a single bond is a single bond regardless of the atoms that form it. These bond fragments are obtained from the off-diagonal elements of the connection table and, like the atom descriptors, they are easily counted and sorted by a digital computer.

The manner in which these descriptors are implemented depends on the specific data set under investigation. In some situations the individual frag-

Table 3.3 Atom Descriptors

1. Total number of atoms	6. Number of fluorine atoms
2. Number of carbon atoms	7. Number of chlorine atoms
3. Number of oxygen atoms	8. Number of bromine atoms
4. Number of nitrogen atoms	9. Number of iodine atoms
5. Number of sulfur atoms	10. Number of phosphorus atoms

Table 3.4 Bond Descriptors

1. Total number of bonds
2. Number of single bonds
3. Number of double bonds
4. Number of triple bonds
5. Number of aromatic bonds
6. Number of delocalized bonds
7. Number of ionic bonds

ment descriptors can be used alone, but in other cases more information can be obtained by combining these descriptors into a smaller subset. For example, in some situations the total number of halogen atoms contained in the compound may be more important than the number of each halogen. In this case the summation of the number of fluorine, chlorine, bromine, and iodine atoms would be a better way of using these particular atom descriptors. In another instance the ratio of the number of oxygen atoms to the number of carbon atoms may be a more useful descriptor. This demonstrates two of the several methods that can be used to combine the fragment descriptors into new descriptors. Naturally, the specific application determines the actual utility of such combinations.

Additional molecular information is indirectly generated along with the general atom and bond information contained within these fragment descriptors. The molecular size and weight are directly related to the number of atoms and bonds in the molecule. Furthermore, if the total number of bonds is greater than or equal to the total number of atoms, a ring system is indicated. The number of hydrogen atoms is implied by the number of unsaturated bonds and the atom types present.

One drawback to the use of fragment descriptors is that most of the structural information is lost. Generally fragment descriptors are of little utility unless additional structural information can be provided. These descriptors are excellent where gross composition is a greater factor than specific orientation, or as complements to other descriptors in which structural information is retained but specificity for the types of connectivity and composition is obscured.

Substructure Descriptors

Information on the presence or absence of specific substructures or functional groups can be very useful in describing molecules. If a substructure is found in a molecule, the descriptor can be given a value equal to the number of times it appears in the structure. If the substructure is not present, the descriptor has a value of zero. To generate substructure descriptors for a given data set, both a fast substructure searching algorithm and an appropriate library of substructures are needed.

In general, substructure searching algorithms fall into two categories. The first, atom-by-atom searching, is the easiest to implement on a digital computer because it simply matches the structure and substructure atoms and associated bonds one at a time using all possible combinations until all matches are found (9). For substructures with one or two atoms, this type of search is very fast. However, an increase in the size of the substructure causes a factorial increase in the number of possible combinations and a corresponding increase in the time it takes for a single search. Consequently, this type of searching is done only when a few structures need to be searched.

The second category of substructure searching algorithms utilizes set reduction techniques to either avoid or vastly reduce the number of factorial calculations involved in any given search. Although they are more complex than atom-by-atom searches, algorithms implementing set reduction techniques are very attractive because of their faster searching speeds. To explain how set reduction operates, a brief introduction is presented.

The first step in a set reduction substructure search is to generate a series of atom subsets for both the substructure and structure according to a set of properties. This is illustrated in Table 3.5, where a series of atom subsets has been generated for three substructures and a structure using nine structural properties. The presence of an empty subset of structural atoms for a corresponding nonempty substructure subset is sufficient proof that the structure does not contain the substructure. Therefore, since the structure does not contain any nitrogen atoms, substructure I cannot be in the structure and the search can be terminated for this particular substructure. However, further processing must be carried out for the remaining two substructures before a conclusion can be reached.

The next step in the searching process is referred to as partitioning, which attempts to find common elements of subsets by set intersection (logical AND). For substructure III in Table 3.5 this is performed by the intersection of subsets corresponding to lines 2, 4, and 8. When this is done, the result is an empty set, which is sufficient proof that substructure III is not in the structure. Therefore, the search procedure can be terminated for this substructure.

However, this is not the case for substructure II. The partitioning of the subsets corresponding to lines 1, 4, 5, and 9 for substructure atom a and lines 2, 5, and 7 for substructure atom b produced two new subsets, which are shown at the bottom of Table 3.5. Since a one-to-one correspondence was

Table 3.5 Some Properties and Atom Subsets for Demonstrating Set Reduction Substructure Searching

Set Generating Property	Substructure I $z-\underset{p}{N}-y$ Subsets	Substructure II $c-\underset{a}{\overset{\overset{b}{\underset{\|}{O}}}{C}}-d$ Subsets	Substructure III $y-\underset{x}{O}-z$ Subsets	Structure $\underset{1}{CH_3}-\underset{2}{\overset{\overset{O}{\|}}{C}}-\underset{4}{CH_2}-\underset{5}{CH_2}-\underset{6}{\overset{\overset{_8CH_3}{\|}}{C}}=\underset{7}{CH_2}$ Structure Subsets	Line
Carbon atoms	∅	[a]	∅	[1, 2, 4, 5, 6, 7, 8]	1
Oxygen atoms	∅	[b]	[x]	[3]	2
Nitrogen atoms	[p]	∅	∅	∅	3
Forms single bonds	[p]	[a]	[x]	[1, 2, 4, 5, 6, 8]	4
Forms double bonds	∅	[a, b]	∅	[2, 3, 6, 7]	5
Forms triple bonds	∅	∅	∅	∅	6
Has one connection	∅	[b]	∅	[1, 3, 7, 8]	7
Has two connections	[p]	∅	[x]	[4, 5]	8
Has three connections	∅	[a]	∅	[2, 6]	9
Partition:	[a] [b]			[2, 6] [3]	

77

found for substructure atom b, the only remaining task is to determine the proper assignment for atom a. This can be carried out by simply checking the connectivity around atom 3 in the structure and determining whether atom 2 or 6 corresponds to substructure atom a.

The performance of a set reduction substructure searching algorithm is highly dependent on the properties chosen for subset generation. The more representative these are of structural properties, the faster the algorithm converges on the answer. Another factor governing the performance of set reduction substructure searching is the answer being sought. If only the information on the presence or absence of a substructure is needed, the answer can be determined rather quickly. However, if the number of times the substructure appears in the structure is needed, the search takes longer, especially when the substructure is present.

Several different algorithms have been described that use set reduction for the retrieval of chemical structures with specific substructures (e.g., see references 10–12). To generate substructure descriptors for structure–activity studies, a variation of the technique described by Sussenguth (10) was used. The modifications allowed for greater substructure specificity, a wider variety of substructure types, and numeric search results, all of which are necessary for substructure descriptor generation. A discussion of the changes made in the Sussenguth's algorithm has been previously reported (13) and is not detailed in this book.

The problem of creating a substructure library is not as easy to solve as obtaining a substructure searching algorithm. One approach to this problem involves the systematic combination of the basic atom and bond fragments into substructures. However, the final number of substructures generated in this manner would be totally unmanageable. The discrimination between useful and useless substructures would constitute an enormous amount of data processing. Thus this approach is not a viable alternative. A more workable approach to the problem is to assemble a library of substructures that cover the basic functional groups encountered in the data set, as well as other substructures that, for one reason or another, are thought to be important for the analysis of the data set under investigation.

The actual information contained in any one substructural descriptor depends highly on the judgment of the person selecting the substructure library. In some applications, good descriptors can be obtained immediately because sufficient *a priori* knowledge exists. However, a trial-and-error procedure may be warranted in cases where a large number of possible substructures are generated and poor descriptors are eliminated by some prescreening criterion. In general, substructure descriptors are very important because they restore a portion of the structural information lost in the atom and bond fragmentation. Nevertheless, considerable structural information is still missing.

Environment Descriptors

Fragment and substructure descriptors indicate the components of a molecule, but the manner in which these individual parts are connected is missing. However, environment descriptors take into account how different areas of a molecule fit together and provide a measure of the "environment" in which a single atom fragment finds itself.

The environment descriptor describes the fragment's surroundings by including its first and second nearest neighbors and their bonds in a single parameter that reflects the atom and bond types connected to it. There may be identical fragments in a molecule, but they do not necessarily belong to the same functional group. For example, the fragment $-\overset{\overset{\Vert}{}}{C}-$ is found once in both structures A and B below but twice in structure C. Obviously, the environment seen by this fragment is different in each of these three cases. However, whether the difference can be reflected in the descriptor depends on the procedure used for calculating the parameter.

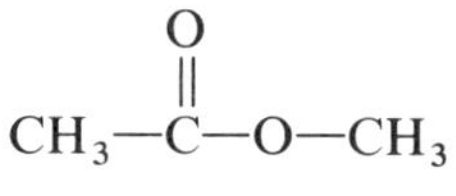

$$CH_3-\overset{\overset{O}{\Vert}}{C}-O-CH_3 \qquad \underset{CH_3}{\overset{CH_3}{\diagdown}}C=CH-CH_3$$

(A) 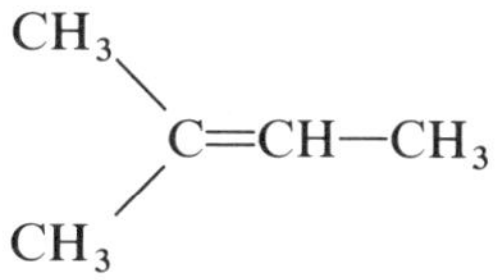   (B)

$$CH_3-\overset{\overset{O}{\Vert}}{C}-CH_2-CH=C\underset{CH_3}{\overset{CH_3}{\diagup}}$$

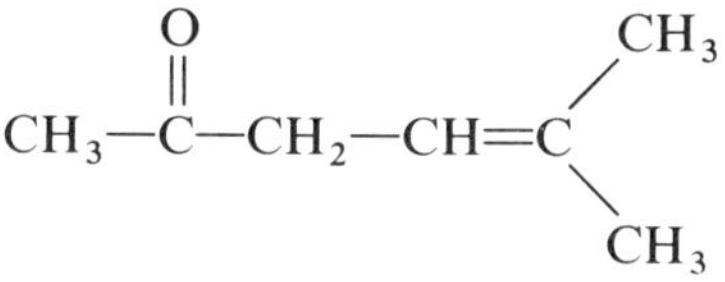

(C)

The three forms of environment descriptors are bond environment descriptors (BED), weighted environment descriptors (WED), and augmented environment descriptors (AED). The BED descriptor is the simplest of the three environment descriptors to calculate. It is obtained by simply summing the number of nonhydrogen bonds for the environmental fragment's first and second nearest neighbors. For example, structure A, which is shown above, contains the fragment $-\overset{\overset{\Vert}{}}{C}-$ only once. This fragment's nearest neighbors consist of a carbon atom with one single bond, an oxygen atom with one double bond, and an oxygen atom with two single bonds. The second nearest neighbor consists of a carbon atom with one single bond. Thus, a value of 5 is obtained for the BED descriptor for this example $(1 + 1 + 2 + 1 = 5)$. To calculate the WED descriptor, the type of bonds connected to the nearest neighbors are taken into consideration, that is, a single bond is 1, a double bond is 2, a triple bond is 3, and an aromatic bond is 4. Therefore, using the same example as above, a value of 6 is obtained for the WED descriptor (the oxygen atom with the double bond would now have a value of 2). Using the same procedure and environmental fragment as above, the BED and WED descriptors for structure B are 5 and 6 respectively, while for structure C a value of 12 is obtained for the BED descriptor and 15 for the WED.

In the calculation of the augmented environment descriptor, both the type of bond and the atoms that form the bond are taken into consideration. First, each atom and bond type is assigned a value. In this work the numeric values employed in the connection table were used for both the atoms and the bonds (see Table 3.2). A value is then calculated for each nearest neighbor by summing together the product of the bond type and the numeric value for each atom for each bond. Thus a carbon–carbon single bond would have a value of 1, a carbon–oxygen double bond would have a value of 4, and a carbon–oxygen single bond would have a value of 2. The descriptor is finally calculated by adding together the values obtained in the above procedure for the first and second nearest neighbors. For structure A, the AED for the example fragment would be equal to 11 (1 from the carbon–carbon single bond, 4 from the carbon–oxygen double bond, 4 from the two carbon–oxygen single bonds, and 1 from the final carbon–carbon single bond). The AED for structures B and C are 6 and 17, respectively.

Since there may be more than one fragment present, the environment descriptor indicates the sum of all the environments for a given fragment. This feature makes them useful when used in conjunction with substructure descriptors. The substructure descriptors indicate the number of times a particular fragment is found in the molecule and the environment descriptors indicate the context in which the fragment is found.

The creation of a fragment descriptor library can best be handled using intuition, chemical knowledge, and experience, as was done in the case of the substructure library. Searching for a given fragment in a structure requires a simple atom by atom search. When the fragment is found, the calculation of the environment descriptor is easily accomplished using the connection table information.

The concept of the environment is not limited to connectivities, but can take into account electron densities, bond distances, electronegativities, or other physical parameters. This can be done by replacing the values assigned to the bond and atom types by the desired parameters. In this manner more informative descriptors can be obtained.

Use of the environment descriptors may reveal relations that are not particularly obvious. Note that both structures A and B above have the same BED and WED values. These structures, which at first glance appear quite different, do indeed have these parameters in common. However, when the type of atoms connected to these bonds is taken into account, the difference becomes apparent. Such relationships may or may not prove significant. Their utility depends on the type of environment measured, the molecule being coded, and the problem being attacked.

Molecular Connectivity

The representation of a molecule in the form of a connection table provides more than a convenient form of molecular storage. The connection table represents the topology of the molecule. From this representation, it is possible to derive descriptors that are related to several fundamental physical phenomena.

Another useful topological descriptor is the branching index, which was originally developed by Randić (14) to provide a topological index that could characterize the amount of branching in hydrocarbon molecules. Randić demonstrated several interesting correlations between this branching index and the boiling points of alkane isomers, surface areas of selected saturated acyclic hydrocarbons, enthalpies of formation of alkane isomers, and correlations between the parameters of the Antoine equation, which relates the vapor pressure of hydrocarbons to the temperature. The branching index can also be used to develop an index similar to that of the Kovats index for branching. The Kovats index was developed using normal alkanes as standards to define a scale of retention values. The index for a substance is then obtained by measuring the position of the retention (on a logarithmic scale) relative to the nearest standard. The Kovats index is therefore an empirically derived index. It can be shown that the branching index develops correlations identical to those of the Kovats index, the scale differing only by a factor of 200 (13).

Applications of the branching index to studies of structure–activity relations have been carried out by Kier and co-workers. They have found significant correlations between the branching index and solvent cavity surface area, molecular polarizability, local anesthetic potency of a molecule, water solubility, boiling point, and the partition coefficient of a molecule. Also noted has been the utility of the branching index in correlations with the biological activity of a whole spectrum of molecules, and even the ability to define parabolic relationships between the connectivity index and biological activity. More recently, a generalized theory for the development of molecular connectivity has been worked out by Kier and Hall and has been published in book form (15).

On the basis of the above work, it would appear that the branching index codes some very fundamental properties of molecular structure. This is not very surprising, since molecular properties such as boiling point, vapor pressure–temperature relation, free energy, heat of solution, density, molecular volume, and refractive index are known to be dependent on chemical structure. The branching index is seen to provide some basic information concerning the overall composition of the molecule.

The calculation of this molecular connectivity descriptor is straightforward (15) and easy to do using the connection table information. First, each nonhydrogen atom (i) is assigned a value (L_i) corresponding to the number of nonhydrogen atoms connected to it. To take into account the different types of bonds that can connect atoms, one unit is added to L_i for each pi bond atom i forms. Therefore, $L = 2$ for $CH_2=$, $L = 3$ for $CH\equiv$, and $L = 4$ for $R_2C=$. This addition has been found to be necessary in studies where the branching index is being related to partition coefficients (15). Once each atom has been assigned a value, a number is derived for each bond in the structure by calculating the product of the numbers associated with the two atoms forming the bond. The reciprocal of the square root of this number is then calculated and becomes the bond value (C_k). Finally, the connectivity index for the entire molecule is derived by the summation of all C_k values for all of the bonds in the molecule. The general index is then used as one descriptor.

Since a cyclic compound has one more bond than the corresponding straight chain isomer, a correction for rings is required. This correction is accomplished by subtracting from the overall index a value equal to the average contribution of all bonds in the ring system. This corrected value is then used as a second molecular descriptor.

Another index can be calculated by including information about the type of atoms forming each bond (this is similar to the augmented environment descriptor). In this way more information is encoded about the overall chemical nature of the molecule rather than merely about the amount of branching. Other variations of the molecular connectivity have been proposed by Kier and Hall and can be found in their book (15).

The calculation of these descriptors is a straightforward procedure that is easily programmed for a digital computer. The connection table contains all the information needed to calculate these molecular connectivity descriptors. As would be expected, the amount of computer time involved in the calculation of these indices is extremely small. Although this branching index has been correlated to a large variety of different properties, its utility in studies of structure–activity relations undoubtedly depends on the specific problem under investigation and the manner in which it is used.

MOLECULAR MODELING AND GEOMETRIC DESCRIPTORS

The descriptors discussed so far have all been generated from the molecular connection tables representing the two-dimensional structural diagrams of the molecules. However, since molecules are actually three dimensional, it would seem only logical to generate descriptors that incorporate the three dimensionality of the structure.

Unfortunately, obtaining the actual structural shape of any given molecule is not a simple task. The use of x-ray data is not reasonable, since the probability is very small that all the compounds in a given data set have been studied. Other empirical methods could be used to obtain structural information, but this would be extremely costly and time consuming for a data set usable in structure–activity studies. Three-dimensional space filling models could be constructed by hand, but this would be extremely tedious and the extraction of geometric parameters from the models would be difficult at best. Thus there is only one viable alternative—the computer calculation of molecular models using molecular mechanics.

The research area of molecular mechanics deals entirely with the calculation of molecular geometries and energies using classical mechanical principles. Ideally, these calculations should be done using the appropriate Schrödinger equation for the molecular system under investigation. In practice this is not done because of the complexities and computational difficulties encountered. Even though simplified quantum mechanical treatments have been reported, the desired degree of accuracy for large organic molecules has still not been attained. Nevertheless, this method is clearly the preferred approach, and accurate results for large molecules may be attainable in the future. In the meantime, another technique is being applied to calculate geometric models.

A molecule can be viewed as a collection of particles held together by simple harmonic or elastic forces. These forces can be defined by potential energy functions whose terms are the atom coordinates of the molecule. This function can then be minimized to obtain a strain-free three-dimensional model of the molecule. Geometric parameters can then be extracted. A wealth of information already exists describing the procedure and results of several different molecular mechanics algorithms (16–20). Therefore, finding and implementing an algorithm to model sets of molecules is a relatively straightforward procedure. MOLMEC, a modified version of the molecular mechanics routine described by Wipke et al. (21), was developed so that geometric descriptors could be derived for a data set of olfactory stimuli. The basic parts of the system include an input section, a strain minimization section, and an interactive section.

Table 3.6 Bond Length Strain Function:
$E_{bond} = K_b/2 \, (L - L_0)^2$

Bond Type	L_0 (Å)	Bond Type	L_0 (Å)
C—C	1.54	N=N	1.25
C=C	1.34	N≡N	1.10
C≡C	1.20	N—S	1.78
C⋯C	1.39	N=S	1.66
C—O	1.43	N—F	1.36
C=O	1.22	N—Cl	1.79
C—N	1.47	N—Br	1.88
C=N	1.29	N—I	2.07
C≡N	1.16	S—O	1.43
C⋯N	1.34	S=O	1.66
C—S	1.82	S—S	2.05
C=S	1.71	P—O	1.61
C—F	1.34	P=O	1.72
C—Cl	1.74	P—S	1.86
C—Br	1.94	P=S	2.14
C—I	2.12	P—F	1.54
C—P	1.84	P—Cl	2.04
O—O	1.48	P—Br	2.18
N—O	1.36	P—N	1.84
N=O	1.22	P⋯N	1.56
N—N	1.45		

$K_b = 311.9$ for single and aromatic bonds and 500.0 for double and triple bonds, L_0 = expected bond length, and L = observed bond length.

The structure input section of MOLMEC has been designed to allow the user to either read the molecule's connection table from disc storage files or accept the structure from the algorithm UDRAW. Thus MOLMEC can be used independently to look at single molecules or in conjunction with data files to model a large set of compounds. Once the molecule has been entered, control branches to the interactive part where the user can direct the different phases of modeling, as well as monitor the results.

In the strain minimization section, the atom coordinates are systematically altered until a minimum is found in the strain or potential energy function. The actual strain function used in MOLMEC is

$$E_{strain} = E_{bond} + E_{angle} + E_{torsion} + E_{nonbonded} + E_{hybrid} + E_{stereo}$$

Before discussing the individual terms of this function, we should point out that a minimal set of parameters was chosen by Wipke and co-workers so that generality could be obtained. Naturally, some precision was sacrificed to accomplish this. However, the net result has been a molecular modeling routine that produces good models for a large variety of chemical compounds rather than excellent models for a limited series. Since the resulting models from this routine were only going to be used to generate general shape parameters, the need of a function with a high degree of precision was not warranted.

The first two terms of the strain function are for bond stretching and angle bending. The exact equations for these terms can be found in Tables 3.6 and 3.7, respectively. As can be seen, the bond lengths are very general in nature, with no special considerations being made as to where the bond appears in the structure, that is, in an acyclic section, in a ring system, next to an aromatic bond system, and so on. Likewise, the parameters in the bond angle strain term are only the basic terms.

The third term of the strain function relates to torsional bond strain and is given in Table 3.8. The parameters for single bond torsional strain are not typically found in other molecular mechanics functions; here they were chosen to efficiently move eclipsed bonds to the staggered conformation. However, the parameters used for the torsional strain about the other bond types are similar to other functions.

The nonbonded term in Table 3.9 is very basic and corresponds to a hard sphere model of the atom in that no attractive terms are present. The second power term is used in the early stages of modeling to allow atoms to easily

Table 3.7 Bond Angle Strain Function: $E_{angle} = K_a/2 (\Theta - \Theta_0)^2$

Hybrid Type	θ_0 (57.3)	K_a
Sp^3	109.5	80.1
Sp^2	120.0	100.0
Sp	180.0	150.0
Sp^3*	109.5	20.0

K_a = angle strain constant, Θ_0 = expected bond angle, and Θ = observed bond angle.
*Noncarbon atom centers with tetrahedron shaped orbitals.

Table 3.8 Torsional Angle Strain Function: $E_{torsion} = K_t F (\Phi')^2$

Torsional Bond Type	K_t	F	Φ' (57.3)
—A—B—	1.0	1.0	$60 - \Phi, \Phi < 60$
—A=B—	15.0	1.628	Φ (cis)
			$180 - \Phi$ (trans)
=A—B=	0.003	1.628	Φ (cis)
			$180 - \Phi$ (trans)
⋯A⋯B⋯	15.0	1.628	Φ
A—B=C=D—E	15.0	1.6	$90 - \Phi$ (for angle ABDE)

Φ is the measured dihedral angle in degrees.

Table 3.9 Nonbonded Strain Function:
$E_{nonbonded} = K_{nb}/2 \, (D - D_0)^M$

Interaction Type	D_0 (Å)
C—C—C (1,3 nonbonded)	2.52
C—C—heteroatom (1,3 nonbonded)	2.25
All other nonbonded interactions	3.5 (poor models)
	3.0 (good models)
If $D_0 > D$, $E_{nonbonded} = 0$	

$K_{nb} = 28.76$, D_0 = expected interatomic distance, D = observed interatomic distance, $M = 2$ for poor models, and $M = 6$ for good models.

Table 3.10 Hybrid Strain Function:
$E_{hybrid} = 10.0 \, [(ASUM - ASUM_0)/57.3]^2$

If sp^3 carbon, then $ASUM_0 = 380°$ and ASUM is the summation of the four smallest angles around the carbon atom.

If sp^3 carbon and only three atoms attached to the carbon atom, then $ASUM_0 = 330°$ and ASUM is the summation of all three angles.

If sp^3 tetrasubstituted carbon, then $E_{hybrid} = 0$ for ASUM $\geq 380°$.

If sp^3 trisubstituted carbon, then $E_{hybrid} = 0$ for ASUM $\leq 330°$.

If sp^2 carbon, then $ASUM_0 = 330°$ and ASUM is the summation of the three angles around the carbon atom.

If sp^2 carbon, then $E_{hybrid} = 0$ if ASUM $> ASUM_0$.

pass each other to escape initial, bad geometries. The sixth power is used in the final stages because it simulates the hard sphere model better.

The hybrid term was added to the strain function to aid in the transition of molecules in poor initial geometries to good models. This simple term, shown in Table 3.10, merely assures that the desired geometries about sp^2 and sp^3 atoms are indeed favorable. The final term is a special term for asymmetric atoms for which the spatial configuration of its attachments is known. The exact form of this term is given in Table 3.11 where A, X, Y, and Z are the four attachments around the asymmetric center C. This term assures that the asymmetric center maintains the correct relationship with the XYZ plane, which is specified by the user upon entry of the structure. Although not heavily used, this term does play an important role in modeling some molecules.

The actual minimization of the function is best accomplished by some type of nonlinear programming method, such as steepest descent, parallel tangents, and pattern searching. In MOLMEC, an adaptive pattern search routine (22) is used because it is straightforward and requires no analytical derivatives. In a pattern search each atom is moved individually into a low strain position. After each atom is moved a first time, another pass is made to see if any of the atoms can be moved to a more favorable position. This process continues until the overall strain of the molecule does not change, that is, each atom is at its lowest strain. The direction and amount of each move is determined through a feedback process that keeps track of successful moves and makes corrections when moves produce an increase in the overall strain of the molecule.

The amount of time necessary to obtain good molecular models depends on the number of atoms in the molecule, the initial strain of the molecule, and the degrees of freedom in the structure. If a small molecule is being modeled, only one pass through the minimization section of MOLMEC may be sufficient to obtain a good structure. However, this is seldom the case. Usually, the molecules are rather large and require several passes. The actual amount of time per pass is limited by a cutoff parameter in the program so

Table 3.11 Stereochemistry Strain Function:
$E_{stereo} = 0.5 + 100(h - 0.2)^2$

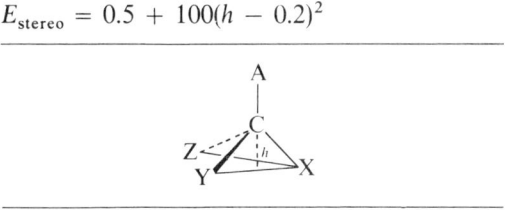

Table 3.12 Staging of Molecular Modeling Routine by Weighting Strain Energy Functions

Stage[a]	I	II	III	IV	V
Strain/atom	>100	50–100	25–50	2.5–25	<2.5
W (angle)	1.0	1.0	1.0	1.0	1.0
W (bond)	0.1	0.2	0.5	1.0	1.0
W (hybrid)	1.0	1.0	1.0	0.0	0.0
W (torsional)	0.0	2.0	2.0	1.0	1.0
W (nonbond)	10.0	3.0	3.0	1.0	1.0
W (stereo)	1.0	1.0	0.0	0.0	0.0

[a] W would be used in the following equation for stage I: $E_{strain}(I) = W(I) \cdot E_{strain}(actual)$.

that the user may analyze the progress of the modeling at different intervals.

A process called staging was added to the program to aid in the rapid calculation of good molecular models. This is accomplished by weighting the energy strain functions by various amounts and changing these weights as a result of the strain per atom value. Table 3.12 shows the weighting factors used for each strain function term.

The graphics interaction section of MOLMEC contains routines capable of rotating and aligning the molecule into any desired position. Since the graphics unit is only a two-dimensional screen, rotation is essential to obtain a good view of the structure. Furthermore, these routines are useful in locating atoms trapped in local minima. If such an atom is found, the user can move the trapped atom to a new position by a MOVE routine found in the graphics section. Naturally, if the structure is altered, the molecule should be passed through the minimization routine at least once more.

When the molecule is finally in a low strain conformation, either the molecular parameters can be listed on an output device or else the structure's coordinate matrix can be stored on a disc file for further processing.

An automatic version of MOLMEC has also been developed so that large molecular data sets can be modeled without continuous supervision. The program consists of an input section, which reads the molecule's connection table and present coordinate matrix from the disc files; a minimization section with all output suppressed; and a section that stores the final coordinate matrix. Good models can easily be obtained in this manner. However, before the coordinate matrices can be used for calculating descriptors, the structures are reviewed to make sure that the molecules are in acceptable conformations.

To get an estimate of the accuracy of this molecular modeling program, four test compounds for which reliable x-ray data exist were modeled. These

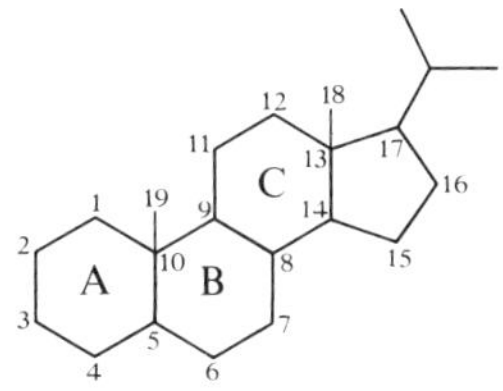

Molecular Modeling and Geometric Descriptors

17-β-Isopropylandrostane

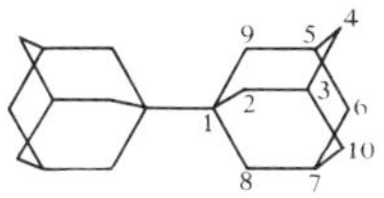

1-Biadamantane

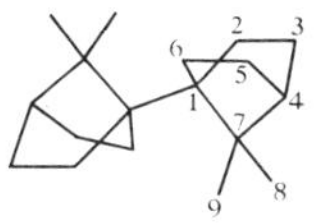

1-Biapocamphane

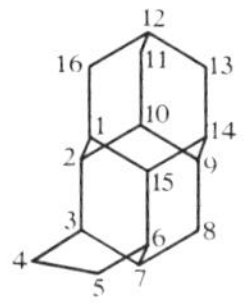

Hexacyclo[10,3,1.0^{2,10},0^{3,7},0^{6,15},0^{9,14}]hexadecane

Figure 3.3 Compounds used to test molecular modeling program's accuracy.

compounds are shown in Figure 3.3 and are the same examples chosen by Altona and Faber (19) to compare other molecular modeling routines. Thus a general comparison of this program with other routines can also be made.

For the first test compound, 17-β-isopropyl androstane, it was found that the average bond length deviation between the modeled and experimentally determined structures was only 0.008 Å, with the largest deviation being only 0.022 Å. This was averaged over 20 bonds in the steroid backbone. The bonds between atoms 2–3 and 3–4 were excluded because the x-ray data were for the compound with a substituent on carbon 3 and the isopropyl group was ignored because of possible crystal lattice distortion. When these results were compared with other molecular modeling routines, no major differences were observed for the bond lengths. However, this trend did not continue into the comparison of the torsional angles. When compared with the experimental data, deviations of 4.7, 4.7, and 5.9° were found for the torsional angles in the A, B, and C rings, respectively. These torsional angle deviations are much larger than any of the other calculated models reported by Altona and Faber. The reason for these poor torsional angles is undoubtedly the hard sphere nonbonded interaction term and the simple single bond torsional angle term used in the strain function. If more accurate models were generated, these two terms would have to be modified.

For 1-biadamantane, very good results were obtained for the carbon–carbon single bonds in the ring system (average deviation = 0.008 Å, largest deviation = 0.011 Å). However, for the single bond connecting the two adamantane sections, a deviation of 0.035 Å was observed. This same trend was also seen for 1-biapocamphane except that the largest deviation was found for the bond between atoms 1 and 7 in the camphane ring. In these two cases, the error can be directly attributed to the nonbonded interaction term. Interestingly, the other molecular modeling routines tested by Altona and Faber also had difficulty with these two compounds. The comparison of the last example compound produced results similar to those obtained for the first three examples.

In general this little experiment has shown that the strain function developed by Wipke and co-workers can produce reasonably good models for compounds that have a rigid structure. The results of the above comparisons are very good considering the simple nature of the strain function used and the fact that no adjustments were made for the difference between modeling a compound as if it were in a vacuum and measuring a compound in a crystal lattice. In the comparison of this modeling routine with the others, no outstanding differences could be seen. Each routine has its weak and strong points. Although more accurate strain functions are needed to do thermodynamic calculations from computer generated models, the function described here does provide good three-dimensional structure representations from which geometric descriptors can be calculated.

At present two basic types of geometric descriptors are calculated from the molecular structures. The molecules' three principal axes of rotation form the basis for the first type of geometric descriptor. These axes are calculated by a radius of gyration program comprised of the following steps:

1. Calculate the center of mass, $\bar{u}$, for the molecule where element u_j is

$$u_j = \frac{1}{M_T} \sum_{i=1}^{N} m_i x_{ij} \quad \text{for } j = 1, 2, 3$$

and M_T is the total mass of the molecule, m_i is the atomic mass of atom i, N is the total number of atoms in the structure, and x_{ij} is the jth coordinate for the ith atom.

2. Calculate the tensor of gyration matrix **R** where the element r_{jk} is

$$r_{jk} = \frac{1}{M_T} \sum_{i=1}^{N} r_i (x_{ij} - u_j)(x_{ik} - u_k) \quad \text{for } j = 1, 2, 3; k = 1, 2, 3$$

3. Diagonalize the tensor of gyration matrix to obtain its eigenvalues.

The actual diagonalization of the tensor of gyration matrix is done by the Jacobi method, since the matrix is symmetrical. The eigenvalues obtained correspond to the three principal radii of the molecule. Since the orientation of the original molecule in space is essentially random, the radii must be sorted in some manner. This is done by arbitrarily assigning X to the longest radius, Y to the second longest radius, and Z to the shortest radius. Once sorted, the three ratios (X/Y, X/Z, and Y/Z) are also calculated. Because of their small values, all the radii are multiplied by some constant scaling factor to prevent loss of information during truncation of the values to integers. These six geometric parameters can then be used as descriptors.

The van der Waals volume of a molecule is the other type of geometric descriptor generated. Before this calculation can be done, the bond distance and the van der Waals radii of the atoms must be known. The bond distances are easily obtained from the molecular modeling results. For the van der Waals radii, an article published by A. Bondi (23) was consulted. The volume occupied by an atom is taken as that of a sphere with radius equal to the van der Waals radius of the atom minus the volume of overlap with the adjacent atoms. The overlap volumes are calculated from standard spherical geometry formulas. However, the true volume is not found because the assumption of sphere and spherical segments is not totally correct, and the radii used were selected as being the "best" values from a large collection of data using an empirical selection method. Table 3.13 contains the van der Waals radii

Table 3.13 Van der Waals Radii Used in the Molecular Volume Calculation

Atom Type	Radius (Å)	X—H ($Å^3$/H atom)
C—	1.70	1.83
C=	1.70	1.83
C≡	1.78	1.36
C⋯	1.77	0.50
O—	1.52	2.29
O=	1.50	—
N—	1.55	2.38
N=	1.55	2.38
N≡	1.60	2.23
N⋯	1.60	2.23
S—	1.80	5.55
S=	1.75	—
F—	1.50	—
Cl—	1.75	—
Br—	1.85	—
I—	1.97	—
P—	1.80	2.86
H—	1.20	
H⋯	1.00	

actually used in the calculation. The total molecular volume is taken as the sum of the contributions for each atom in the structure. The volume contributions of attached hydrogen atoms are also included in the final volume.

To make the routine more versatile, the option of either using standard bond distances or modeled bond distances was included. Since MOLMEC uses the standard bond distances to determine a low strain geometry, it is not surprising that for a well modeled data set, the molecular volumes calculated using the two different bond distances are very similar. However, discrepancies can arise when the molecule contains rings of five or fewer atoms, which cause a large amount of bond strain. The volumes are initially calculated in units of cubic angstroms per molecule, but are later converted to units of cc/mole. The molecular volume can then be used as another geometric descriptor.

Each geometric descriptor contains some information about the molecule. The radii and ratios describe the general shape of the molecule, which may be very important in systems where receptor sites are involved. However, this is only a relative shape, since the model obtained is for the molecule in a vacuum: in some environments, the molecule's shape changes, especially if long chains are present. On the other hand, the molecular volume is essentially constant regardless of how the molecule is bent. However, as for any other descriptor, the actual value of any geometric descriptor depends on the specific application in which it is used.

SUMMARY

The descriptor routines discussed in this Chapter represent several different approaches to extract information from a molecular structure. Each routine requires a different level of computational effort and yields different information about the structure being studied.

Fragment descriptors are easy to compute from the molecular connection tables and reduce molecules to their simplest units. Although structural information is lost, information about the chemical nature of the entire molecule is retained. The amount of unsaturation and the presence of certain heteroatoms are contained in these fragment descriptors and may prove to be useful in some structure–activity studies.

The environment and substructure descriptors are similar in that they both contain information about the molecule's structural makeup that was lost in the fragmentation process. However, they differ considerably in the amount of computational effort necessary to generate the descriptor. For the environmental descriptors, an atom-by-atom search for a single atom fragment followed by a straightforward calculation is all that is necessary to calculate the descriptor. To generate substructure descriptors, a more complex searching algorithm that can handle multiple atom fragments is required. Although generated differently, both of these descriptors carry information about the chemical and structural nature of the molecule.

The molecular connectivity index is an especially informative measurement and one that is easy to compute. Although it is similar to the WED environment descriptor, it contains information about the entire molecule rather than about certain isolated atom fragments. If its utility can be gauged by the number and diversity of its correlations to other physical properties, it is indeed an informative descriptor.

Finally, there are the geometric descriptors that provide a large amount of information about the overall molecular shape and very little about the chemical nature of the molecule. Unfortunately, to obtain geometric descriptors, a molecular mechanics program must be implemented and the data sets members must be modeled. Both steps require a considerable amount of time. However, once the molecules have been modeled and the coordinates calculated, the generation of the descriptors can be done very quickly.

As can be seen from the preceding discussion, the variety of molecular descriptors is wide. The descriptors used for attacking a particular problem are dependent on the biological or physical property being studied. There appear to be no limits on the types of descriptors that could be generated. It is in the design of new ways to adequately describe molecular structure that much of the progress in structure–activity studies remains to be made. It is here that the creativity and insight of the scientist can be brought to bear most directly on the problem. The future should see the development of many new, more useful ways to capture the essence of molecular structure in strings of numerical descriptors for structure–activity studies.

REFERENCES

1. W. J. Wiswesser, *A Line-Formula Chemical Notation*, Thomas Y. Crowell Co., New York, 1954.
2. E. G. Smith, *The Wiswesser Line-Formula Chemical Notation*, McGraw-Hill, New York, 1968.
3. C. H. Davis and J. E. Rush, *Information Retrieval and Documentation in Chemistry*, Greenwood Press, Westport, Conn., 1974.
4. M. F. Lynch, J. M. Harrison, and W. G. Town, *Computer Handling of Chemical Structure Information*, Macdonald, London, 1971.
5. D. J. Gluck, A Chemical Structure, Storage and Search System Designed at DuPont, *J. Chem. Doc.*, **5**, 43 (1965).
6. H. L. Morgan, The Generation of a Unique Machine Description for Chemical Structures—A Technique Developed at Chemical Abstracts Service, *J. Chem. Doc.*, **5**, 107 (1965).

7. W. E. Brugger and P. C. Jurs, Molecular Structure Input Program Using a Storage Cathode Ray Tube Terminal, *Anal. Chem.*, **47**, 781 (1975).
8. W. E. Brugger and P. C. Jurs, UDRAW (Program No. 300), Quantum Chemistry Program Exchange, Department of Chemistry, Indiana University, Bloomington, Ind., 47401.
9. L. C. Ray and R. A. Kirsch, Finding Chemical Records by Digital Computers, *Science*, **126**, 814 (1957).
10. E. H. Sussenguth, A Graph-Theoretic Algorithm for Matching Chemical Structures, *J. Chem. Doc.*, **5**, 36 (1965).
11. T. K. Ming and S. T. Tauber, Chemical Structure and Substructure Search by Set Reduction, *J. Chem. Doc.*, **11**, 47 (1971).
12. J. Figeras, Substructure Search by Set Reduction, *J. Chem. Doc.*, **12**, 237 (1972).
13. G. S. Zander and P. C. Jurs, Generation of Mass Spectra Using Pattern Recognition Techniques, *Anal. Chem.*, **47**, 1562 (1975).
14. M. Randić, On Characterization of Molecular Branching, *J. Am. Chem. Soc.*, **97**, 6609 (1975).
15. L. B. Kier and L. H. Hall, *Molecular Connectivity in Chemistry and Drug Research*, Academic, New York, 1976.
16. J. E. Williams, P. J. Strang, and P. von R. Schleyer, Physical Organic Chemistry: Quantitative Conformational Analysis; Calculation Methods, *Ann. Rev. Phys. Chem.*, **19**, 531 (1968).
17. A. J. Hopfinger, *Conformational Properties of Macromolecules*, Academic, New York, 1973.
18. E. M. Engler, J. D. Andose, and P. von R. Schleyer, Critical Evaluation of Molecular Mechanics, *J. Am. Chem. Soc.*, **95**, 8005 (1973).
19. C. Altona and D. H. Faber, Empirical Force Field Calculations. A Tool in Structural Organic Chemistry, *Top. Curr. Chem.*, **45**, 1 (1974).
20. N. L. Allinger, Calculation of Molecular Structure and Energy by Force-Field Methods, in *Adv. Phys. Org. Chem.*, Vol. 13, V. Gold (Ed.), Academic, New York, 1976.
21. W. T. Wipke, T. M. Dyott, and J. G. Verbalis, Abstracts, 161st National Meeting, American Chemical Society, Los Angeles, Calif., March 1971.
22. E. S. Buffa and W. H. Taubert, *Production-Inventory Systems, Planning, and Control*, R. D. Irwin, Inc., Homewood, Ill., 1972.
23. A. Bondi, Van der Waals Volumes and Radii, *J. Phys. Chem.*, **68**, 441 (1964).

CHAPTER 4

Pattern Recognition: Linear Discriminant Functions

In Chapter 2 it is seen that the underlying assumption of pattern recognition is that "similar" objects tend to cluster in limited regions of the space formed from their descriptors. The object of pattern recognition analysis is the development of discriminant functions that will define the boundaries between these clusters. Additionally, Chapter 2 discusses two methods of developing such functions—parametric and nonparametric.

Parametric methods derive the form of the optimal discriminant function by approximation of the form of the underlying distribution function. Some form of the Bayes relation is then used to develop a classification rule. Because of the computational advantages realized, a large number of parametric methods make the assumption that the data are normally distributed. Other assumptions often used include equal covariance matrices, identical class means, equal loss functions, equal class conditional probabilities, and equal *a priori* class probabilities. The effect of such approximations on the accuracy of the decision function is highly problem dependent.

Nonparametric methods make assumptions concerning the form of the optimal distribution function and then use any one of several methods to fit this function to the data. Linear decision functions are those used most often because they are easily dealt with computationally and they provide a reasonable approximation. Nonparametric algorithms do not ignore the underlying distribution of the data. The distribution is implicitly used when attempting to place the decision surface in an optimal position.

Let us not overlook the fact that optimal is as defined by the Bayes relation. Certainly if we knew the Bayes parameters exactly we could directly calculate the discriminant function that has the minimum classification error. Although it is often forgotten, both parametric and nonparametric methods are approximations to this ideal. Parametrics are approximate because of their simplifying assumptions concerning the statistical characteristics of the data. Nonparametric methods are approximate because of their simplifying assumptions concerning the form and the fit of the optimal discriminant function.

The question becomes one of which approach is most effective in the development of structure–activity relations. We believe that the nonparametric methods offer an approach more suited to the requirements involved in structure–activity applications. The strongest objection we have to many of the parametric methods is the assumption that the data are normally distributed. There seems little evidence that this approximation will not yield discriminants that deviate severely from those that would be obtained if the Bayes relation could be utilized. Additionally, there seem to be several problems inherent in approximating parameters such as $P(W_i)$ and $P(\mathbf{X}_i)$ directly from biological data. The most difficult problem in studies of structure–activity relations is the determination of which compounds to include as members of the inactive set. In work involving a series of homologues the problem is somewhat alleviated; however, for structurally diverse data, this is not the case. Since active compounds are the exception rather than the rule, the approximation of the above parameters is difficult. Certainly it is unacceptable to assume the data fit a normal distribution for no better reason than it is convenient to calculate.

Other estimations seem especially difficult. How many active compounds are there for every inactive compound? Can we obtain a reasonable estimation of this number from our initial data set? Certainly the manner in which the data are distributed is informative, but in the early stages of analysis are the number of data points sufficient to provide reliable distributional approximations? The complexity, as well as the cost of making these estimations, lessens the appeal of parametric methods. Furthermore, there is no reason to expect distributions that are not unimodal. Such distributions are not conveniently dealt with parametrically. Methods that avoid distributional assumptions seem to offer more promise in providing reasonable classification performance.

Nonparametric methods appear to offer a flexible method of dealing with the development of structure–activity relations. Linear decision functions are easy to use and the error introduced by this approximation appears to be less than that afforded by an incorrect distributional approximation. An additional feature of the nonparametric methods is their ability to adapt to changes in the structure of the data. Such changes can be expected as new insight into the problem is obtained. The next few sections give an overview of the application, capabilities, and limitations of selected nonparametric methods of discriminant development.

THE LINEAR LEARNING MACHINE

One of the simplest of the nonparametric algorithms is the linear learning machine. The linear learning machine is an iterative algorithm that uses error

The Linear Learning Machine

correction feedback to develop a useful linear decision surface. How to express data in a matrix form is shown in Chapter 2. Thus each member of the data set can be viewed as a point in space whose distribution is governed by the components of its description vector. If the components are sufficiently informative, the data cluster into classes that relate to the presence or absence of a particular property. These clusters are often referred to as the yes class and the no class. The object is to develop a surface that bisects the data set such that all members of the yes class lie on one side of the surface, while all members of the no class lie on the opposite side of the surface. As the learning machine is a nonparametric classifier, the assumption is made that a linear surface is the optimal surface with which to effect the classification. The problem is to determine the position of a linear surface at which a reasonable performance is obtained. This requires a method to determine on which side of the plane a point lies. This can be accomplished by first adding one extra dimension to each member of the entire data set and then calculating the dot product between each point and a vector normal to the surface of the plane. A simple example is given below.

Figure 4.1 shows a one-dimensional data set that has been augmented with one extra component. The y axis represents the values for the descriptor and the x axis represents the magnitude of the added dimension. This magnitude is donated by the symbol DPO. For clarity we have imposed on the x axis at the value of DPO a distribution of descriptor values for the two classes. Note

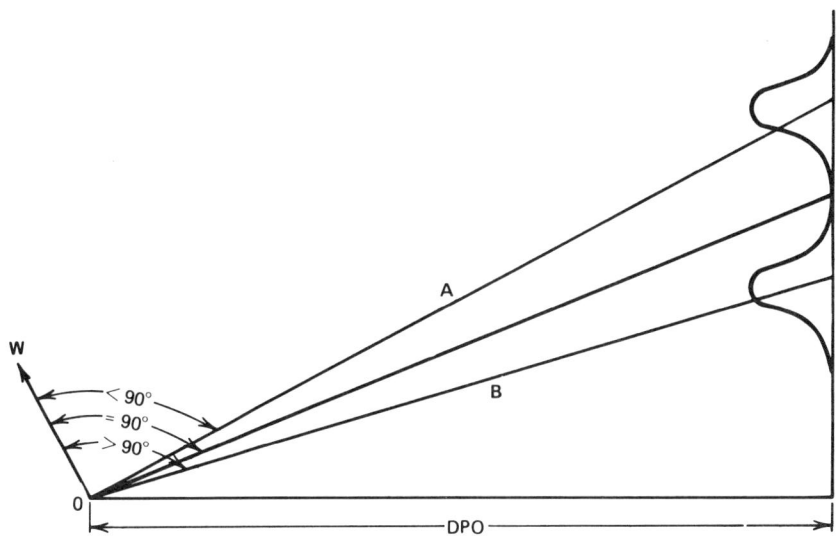

Figure 4.1 Example of a linear separable one-dimensional data set.

that a plane can pass between both of these classes, totally separating them. Data sets having this property are called linearly separable sets. Note also that the decision surface and the two data vectors **A** and **B** pass through the origin. Clearly, if an extra component was not added to the system, the plane would be unable to pass through the origin and separate the sets. As is made clear shortly, the ability of the plane to simultaneously pass through the origin and separate the data is of utmost importance.

The side of the decision surface is defined by constructing a unit length vector, **W**, normal to the surface of the plane. The side of the plane on which a data point lies can now be determined from the value of the dot product of the weight vector **W** with each of the vectors in the data set. One form for the dot product is

$$\mathbf{W} \cdot \mathbf{A} = |\mathbf{W}| |\mathbf{A}| \cos \theta \qquad (4.1)$$

For data to be on the same side of the surface as the weight vector they must lie within 0 to 90° of the weight vector. The dot product is therefore positive. Alternately, data on the side opposite the weight vector have a negative dot product as they must lie within 90 to 180° of the weight vector. Since only the fact that the yes class points are on one side of the plane when the no class points are on the other is of interest, the initial orientation of the weight vector is irrelevant. Therefore, an algorithm that develops weight vectors satisfying these dot product conditions would also define a linear surface that separates the data.

Given that a separable condition exists, a useful surface can be developed by an operation termed training. To train a decision surface the members of the data set are taken sequentially and the dot product is calculated. Whenever any member is misclassified, the weight vector is modified such that the point is correctly classified. This correction process is called negative feedback and continues until all the members of the training set are correctly classified. If convergence is not obtained after some preset number of feedbacks, the algorithm is terminated.

Several methods of feedback that guarantee convergence on linearly separable sets are detailed in the literature (1). One of the simplest and most effective methods is to move the decision surface along a perpendicular axis between the point and the plane so that after correction it is the same distance on the correct side of the point as it was previously on the incorrect side. This operation is performed on the weight vector. Thus if

$$\mathbf{W} \cdot \mathbf{X}_i = S_i \qquad (4.2)$$

is of the incorrect sign for classifying $\mathbf{X}_i$, a new vector, $\mathbf{W}'$, is desired such that

$$\mathbf{W}' \cdot \mathbf{X}_i = S_i' = -S_i \qquad (4.3)$$

The Linear Learning Machine

We wish to develop this vector by combining a fraction, c, of $\mathbf{X}_i$ with $\mathbf{W}$ to form the new weight vector; that is, we wish a new $\mathbf{W}$, $\mathbf{W}'$ such that

$$\mathbf{W}' = \mathbf{W} + c\mathbf{X}_i \tag{4.4}$$

Combining equations 4.3 and 4.4 yields

$$S'_i = \mathbf{W}' \cdot \mathbf{X}_i = (\mathbf{W} + c\mathbf{X}_i) \cdot \mathbf{X}_i \tag{4.5}$$

which can be solved for c to yield

$$c = \frac{S'_i - S_i}{\mathbf{X}_i \cdot \mathbf{X}_i} \tag{4.6}$$

As it is desired that $S' = -S$, c becomes

$$c = \frac{-2S_i}{\mathbf{X}_i \cdot \mathbf{X}_i} \tag{4.7}$$

and the new weight vector becomes

$$\mathbf{W}' = \mathbf{W} - \left[\frac{2S_i}{\mathbf{X}_i \cdot \mathbf{X}_i}\right]\mathbf{X}_i. \tag{4.8}$$

Generally the weight vector is developed using only part of the data set, leaving the remaining members to be used in testing the predictive ability of the classifier. The classification procedure is to form the dot product of the weight vector with each member of the prediction set. The dot product is then used in the following classification rules:

1. If $S_i > 0$, then $\mathbf{X}_i$ is in class 1.
2. If $S_i \leq 0$, then $\mathbf{X}_i$ is in class 2.

Often the performance of the classifier can be improved by using a deadzone, d. The feedback algorithm for the learning machine now corrects the weight vector if

$$a(S_i + d) < 0; \quad a = 1 \text{ for class 1}$$
$$a = -1 \text{ for class 2}$$

The deadzone tends to better align the plane between the classes. The classification rules for prediction become:

1. If $S_i > d$ then $\mathbf{X}_i$ is in class 1.
2. If $S_i < -d$ then $\mathbf{X}_i$ is in class 2.
3. If $-d \leq S_i \leq d$, then $\mathbf{X}_i$ is not classified.

The linear learning machine has several drawbacks. The algorithm has no capacity to optimally terminate if the data are inseparable. Therefore, linearly inseparable data sets cannot be treated adequately. If the set is separable, there is no guarantee that feedback methods will achieve that separation in a reasonable time, only that ultimately the separation will be effected. Lastly, the feedback algorithm simply searches for a position in which the plane separates the data. There are an infinite number of planes that will pass through such a region. Therefore, the surface actually developed may bear little resemblance to one derived parametrically.

Despite these features the linear learning machine offers a viable method of developing a usable decision surface. It is rapid, easy to use, and normally provides excellent discrimination abilities.

GRADIENT DESCENT AND THE LEAST SQUARES ALGORITHM

Although the linear learning machine converges when the classes under consideration are separable by a linear surface, in nonseparable situations the learning machine oscillates for as long as it is allowed to execute. It seems likely that a different method of positioning the decision surface would lead to an algorithm that does not suffer from this drawback. If, for instance, the positioning of the surface was implemented through minimizing or maximizing some criterion, then it might be possible to extend the utility of such positioning algorithms to nonseparable sets. Gradient descent techniques offer an alternate approach to the development of a wide variety of decision functions, certain of which have utility in nonseparable cases.

The basic approach to the gradient descent technique is to form a criterion function that reaches a minimum value when the optimal position of the decision surface is realized. The actual algorithm that results from this approach is dependent on the form of the criterion function. For example, minimization of the criterion function

$$F(\mathbf{W}, \mathbf{X}) = \tfrac{1}{2}(|\mathbf{W}'\mathbf{X}| - \mathbf{W}'\mathbf{X}) \tag{4.9}$$

with respect to $\mathbf{W}$ leads to an algorithm that is essentially that of the learning machine.

As seen in the derivation of the linear learning machine, the act of positioning a surface is effected by recursively altering the weight vector such that a solution is effected. We can express this in the form of a gradient descent procedure by writing

$$\mathbf{W}(k+1) = \mathbf{W}(k) - c\left[\frac{\partial F(\mathbf{W}, \mathbf{X})}{\partial \mathbf{W}}\right]_{\mathbf{W}=\mathbf{W}(k)} \tag{4.10}$$

Gradient Descent and the Least Squares Algorithm

Note that the new weight vector is updated by a factor related to the gradient of the criterion function. When $\partial F/\partial \mathbf{W} = 0$ the function is at a minimum and no more changes are made on the weight vector.

The gradient for equation 4.9 is simply

$$\frac{\partial F(\mathbf{W}, \mathbf{X})}{\partial \mathbf{W}} = \tfrac{1}{2} [\text{sgn}\,(\mathbf{W}', \mathbf{X}) - \mathbf{X}]$$

where

$$\text{sgn}\,(\mathbf{W}', \mathbf{X}) = \begin{cases} 1 & \text{for } \mathbf{W}'\mathbf{X} > 0 \\ -1 & \text{for } \mathbf{W}'\mathbf{X} \leq 0 \end{cases} \quad (4.11)$$

Substituting into equation 4.10 yields

$$\mathbf{W}(k+1) = \mathbf{W}(k) + \frac{c}{2}[\mathbf{X}(k) - \mathbf{X}(k)\,\text{sgn}\,(\mathbf{W}', \mathbf{X}(k))] \quad (4.12)$$

On expanding as per the definition of sgn

$$\mathbf{W}(k+1) = \mathbf{W}(k) + c\mathbf{X}(k) \quad (4.13)$$

which is identical to the learning machine algorithm.

Of course, we still lack a method of intelligently terminating the algorithm when the data are not linearly separable. This example, however, does show how the gradient technique generalizes many of the iterative methods.

An algorithm that can be used for nonseparable cases can be derived by using gradient descent techniques and a judicious choice of criterion functions. We have shown that the learning machine computes a solution to the problem of finding a surface such that

$$\mathbf{X} \cdot \mathbf{W} > 0 \quad (4.14)$$

However, an equivalent way of stating this problem is

$$\mathbf{X} \cdot \mathbf{W} = \mathbf{b} \quad (4.15)$$

where X is the data matrix whose rows represent objects and whose columns represent the corresponding measurements. This matrix is augmented by adding a unit valued component to each of the objects. Additionally, all the members of the no class are multiplied by -1.[1] We can now consider the criterion function

$$F(\mathbf{W}, \mathbf{X}, \mathbf{b}) = \tfrac{1}{2} \sum_{j=1}^{N} (\mathbf{W}'\mathbf{X}_j - \mathbf{b}_j)^2 = \tfrac{1}{2} \|\mathbf{XW} - \mathbf{b}\|^2 \quad (4.16)$$

---

[1] These mysterious operations merely (1) augment the data space with one extra dimension, as explained in the discussion concerning the learning machine. (2) the actual definition for equation 4.15 is $\mathbf{W} \cdot \mathbf{X} = \mathbf{b}$ for class 1 (yes class), and $\mathbf{W} \cdot \mathbf{X} = -\mathbf{b}$ for class 2 (no class). If we multiply all the members of class 2 by -1, then we can write one equation that serves for both classes.

where N is the number of members in the set and $\|\mathbf{XW} - \mathbf{b}\|$ is the magnitude of the vector $(\mathbf{XW} - \mathbf{b})$. This criterion function clearly is at a minimum when equation 4.15 is satisfied. If we minimize this function with respect to both $\mathbf{W}$ and $\mathbf{b}$ we obtain an algorithm that minimizes the sum of the squared error. Such procedures are generally referred to as least mean square error procedures (LMSE).

The minimization of equation 4.16 leads to a classification algorithm that is often referred to as the Ho-Kashyap algorithm. The gradient of this function with respect to $\mathbf{W}$ and $\mathbf{b}$ is

$$\frac{\partial F}{\partial W} = \mathbf{X}'(\mathbf{XW} - \mathbf{b}) \tag{4.17}$$

$$\frac{\partial F}{\partial \mathbf{b}} = -(\mathbf{XW} - \mathbf{b}) \tag{4.18}$$

To minimize this function with respect to $\mathbf{W}$ we can set $\partial F/\partial \mathbf{b} = 0$ and obtain

$$\mathbf{W} = (\mathbf{X}'\,\mathbf{X})^{-1}\mathbf{X}'\,\mathbf{b} = \mathbf{X}^{\#}\mathbf{b} \tag{4.19}$$

$\mathbf{X}^{\#}$ is called the generalized inverse or pseudoinverse of $\mathbf{X}$. Here $\mathbf{W}$ is the weight vector that yields the minimum error for a particular $\mathbf{b}$. However, this may not be the weight vector that separates the set. The separability also depends on $\mathbf{b}$.

Although equation 4.18 expresses the minimum $\mathbf{b}$ for a given $\mathbf{W}$, equation 4.17 must still be satisfied. Therefore, $\mathbf{b}$ must be constrained to be always positive. This can be accomplished by choosing a new $\mathbf{b}$ such that

$$\mathbf{b}(k+1) = \mathbf{b}(k) + \delta\mathbf{b}(k) \tag{4.20}$$

where

$$\begin{array}{ll} \delta\mathbf{b}(k) = 2\mathbf{c}\,[\mathbf{XW}(k) - \mathbf{b}(k)] & \text{if } \mathbf{XW}(k) - \mathbf{b}(k) > 0 \\ \delta\mathbf{b}(k) = 0 & \text{if } \mathbf{XW}(k) - \mathbf{b}(k) \leq 0 \end{array} \tag{4.21}$$

In equations 4.20 and 4.21 k denotes the iteration index, i is the index for the vector components, and $\mathbf{c}$ is a positive correction increment that is usually 0.5. Equation 4.21 can be stated as

$$\delta\mathbf{b}(k) = \mathbf{c}[\mathbf{XW}(k) - \mathbf{b}(k) + |\mathbf{XW}(k) - \mathbf{b}(k)|] \tag{4.22}$$

where $|\mathbf{XW}(k) - \mathbf{b}(k)|$ is the absolute value of the vector $[\mathbf{XW}(k) - \mathbf{b}(k)]$. From equations 4.19 and 4.20 we obtain

$$\begin{aligned} \mathbf{W}(k+1) &= \mathbf{X}^{\#}\mathbf{b}(k+1) \\ &= \mathbf{X}^{\#}[\mathbf{b}(k) + \delta\mathbf{b}(k)] \\ &= \mathbf{X}^{\#}\mathbf{b}(k) + \mathbf{X}^{\#}\delta\mathbf{b}(k) \\ &= \mathbf{W}(k) + \mathbf{X}^{\#}\delta\mathbf{b}(k) \end{aligned} \tag{4.23}$$

Gradient Descent and the Least Squares Algorithm

If we define the error as

$$e(k) = XW(k) - b(k) \qquad (4.24)$$

the following algorithm is obtained:

$$W(k) = X^{\#} b(1), b(1) > 0$$
$$e(k) = XW(k) - b(k)$$
$$W(k + 1) = X^{\#} b(k + 1)$$
$$b(k + 1) = b(k) + c[e(k) + |e(k)|] \qquad (4.25)$$

Here $|e(k)|$ is the vector whose components are the absolute value of the components of $e(k)$.

An interesting feature of the Ho-Kashyap algorithm is the separability test afforded by the error vector. Note that if all the components of this vector are negative no further corrections are made to the weight vector or the **b** vector. It can be shown that this condition will only occur for systems in which the data are not linearly separable. The fact that such a condition exists, however, does not mean that it can be uncovered in a reasonable amount of time. Therefore, several methods of terminating the algorithm are required.

Termination is warranted when all the components of the error vector become zero. At this point a LMSE solution has been effected. Alternatively, the algorithm can be terminated after a preselected number of iterations.

The major drawback to the Ho-Kashyap algorithm is the requirement that the $X'X$ matrix be inverted. However, this only needs to be done once. Additionally, the algorithm can be quite slow in converging. These computational inconveniences are tempered by the separability test and the provision of a LMSE solution for cases that are nonseparable.

The algorithms presented here comprise only one of the several types of gradient descent LMSE algorithms. Derivations were given to demonstrate the use of the gradient descent and LMSE technique for development of nonparametric classifiers. The number of algorithms that can be derived using such an approach is limited only by the number of reasonable criterion functions that can be developed. For example, rather than employing a function such as equation 4.16, which minimizes the error in equation 4.15, we could minimize the function

$$Q = \sum_{i=1}^{N} [Y_i - F(s_i)]^2 \qquad (4.26)$$

where Y_i is $+1$ for yes class members and -1 for no class members. $F(s_i)$ is a function that depends on the value of the dot product.

A useful algorithm that employs equation 4.26 can be developed by expansion of the dot product in terms of the hyperbolic tangent. A derivation of this algorithm is given elsewhere (2).

Clearly, gradient descent and LMSE techniques provide a wide variety of useful nonparametric algorithms. The ultimate utility of these methods depends on the validity of the assumptions concerning the form of the decision function and the criterion function used to position it.

NEAREST NEIGHBOR CLASSIFICATION

Not all nonparametric algorithms attempt to separate the date with a linear surface. Algorithms such as the nearest neighbor rule attempt to predict class membership based on estimation of the class conditional probability $P(W_i/\mathbf{X})$. The nearest neighbor rule is suboptimal in that the error rate for this algorithm is greater than the Bayes error rate. Fortunately, this error rate cannot be greater than twice the Bayes error, and quite often it is less than that. An extensive literature concerning the algorithm exists. Rather than attempting to reproduce this information we simply provide a rationalization of why the algorithm works. Those wishing further insight into the capabilities and limitations of the nearest neighbor rule can consult references 3 to 8.

The nearest neighbor rule for assigning any unknown $\mathbf{X}$ to a class is simply to assign it to the class in which a majority of its nearest neighbors belong. This is equivalent to assuming that the class conditional probability, $P(W_i/\mathbf{X})$, for the unknown is equal to that of its nearest neighbor, which is a reasonable assumption when the sample density is high. Thus the nearest neighbor rule almost always gives results similar to those obtained with the Bayes rule when the unknown is well separated from the classes other than

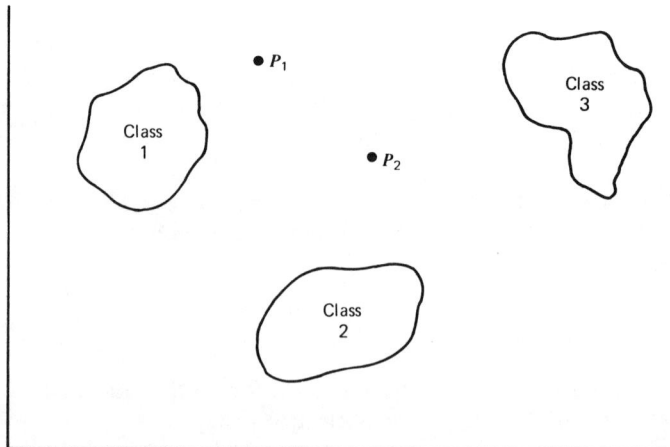

Figure 4.2 Example of the problem inherent in using distance as the classification criterion.

the one to which it belongs. However, they are rarely the same as those obtained with the Bayes rule for cases in which the differences in the probability estimates are quite similar. This is made clear by the example in Figure 4.2, which shows a three-class problem with two unknown points P_1 and P_2. The estimation of class membership is based on the number of neighbors of a particular class that are nearest to the unknown. This estimation is made by calculating the Euclidean distance between the unknown and the known points and then choosing the class according to the unknown's k nearest neighbors. Clearly such a rule will easily result in P_1 being assigned to class 1. Since the measurement of distance does not take into account the probability distribution of each class, the classification for P_2 will not be reliable. The Bayes decision rule would know these distributions and therefore have the lowest probability for error. The nearest neighbor rule could reach the Baysian conclusion only by chance.

From arguments such as these we can intuitively reason that the nearest neighbor algorithm will perform best on data that have well separated distributions or contain a high density of points, and worst in those cases that are highly overlapped or contain a low sample density.

In practice nearest neighbor algorithms perform reasonably well. Any number of neighbors can be used to estimate the class membership. An odd number is always used so as to prevent ties. The algorithm has the advantage of being useful for cases where the data are not separable by a linear decision surface, or in cases where more than two classes are being considered.

The main disadvantage is the requirement that all pairwise distances be computed. This limitation is especially serious for data sets of high dimensionality or a large number of samples. In high dimension spaces a very large number of samples are required for the error bound to approach that of the Bayes classifier. This is because the assumption that $P(W_i/\mathbf{X})$ is equal between nearest neighbors rapidly fails as the density of data points decreases. It can be shown that convergence to the Bayes error rate can be arbitrarily slow and does not even decrease monotonically with the number of neighbors used in the decision process. Of course, this error rate will never be greater than twice the Bayes rate. As with any nonparametric algorithm the degree of success is dictated by the number of samples, number of measurements per sample, and the underlying distribution for each class.

LIMITATIONS OF NONPARAMETRIC LINEAR CLASSIFIERS

An assumption inherent in the use of linear discriminant functions is that the ability to correctly dichotomize the data into classes is meaningful. Thus successful classification is thought to imply that a relationship between the observed properties has been defined. These assumptions can only be tested if certain criteria are met. These criteria deal with the minimum ratio of samples to measurements per sample required to demonstrate a relation within the data. In this section we report the results of investigating these criteria and discuss parameters that indicate the reliability of such relations.

The ability of a function to separate clusters within a set of data is dependent on the dichotomization ability of the discriminant function. Dichotomization ability is the total number of two-class groupings that can be made by a discriminant function. Different functional forms have different dichotomization abilities. The number of two-class groupings that a linear function can effect is given by the following equation:

$$D(N, n) = 2 \sum_{k=0}^{n} C_k^{N-1} \qquad (4.27)$$

where $C_k^{N-1} = (n-1)!/(N-1-k)!k!$, N is the number of samples, n is the number of measurements or variables per sample, and k is an index describing how the groupings are taken.

The total number of dichotomies possible for a set of samples, regardless of its n space distribution[1] is 2^N. Any classifier that could effect all the 2^N possible dichotomies would always indicate that the desired clusters were present regardless of how the data were distributed. This behavior is independent of whether such clusters truly exist. Such a classifier invalidates the assumptions concerning the correlation of properties to measurements made in developing the framework for the pattern recognition system. It would therefore be useless as a classifier.

In cases for which there are more variables than samples, linear classifiers are able to effect all possible dichotomies. However, equation 4.27 indicates that for cases in which there are more samples than measurements, there are fewer linear dichotomies than total dichotomies. Since the number of possible linear dichotomies is dependent on both N and n, the existence of a lower limit to the ratio of samples to measurements seems evident. Below this limit the results of discriminant development are of little use. The probability of randomly assigning class memberships that will yield a relation separable by a linear surface is useful in defining this limit. The form of this probability equation is

$$P = \frac{\text{total number of linearly separable dichotomies}}{\text{total number of dichotomies}}$$

$$= \frac{D(N, n)}{2^N} \qquad (4.28)$$

---

[1] The only assumption made is that the data are well distributed. A data set is well distributed if no subset of $n + 1$ points lies on an $n - 1$ dimensional hyperplane.

Limitations of Nonparametric Linear Classifiers

This equation describes the probability of obtaining a linear dichotomy as a function of the total number of samples, N, and the total number of measurements, or variables, n. When there are fewer samples than variables, a dichotomy is always found regardless of whether it truly exists, that is, $P = 1$. As the number of samples increases with respect to the number of variables, the probability of finding such a dichotomy due to chance decreases, and the value of P decreases.

Linear classifiers are generally implemented by adding one extra dimension to the sample space before training. It is therefore convenient to define a variable λ as being equal to $N/(n - 1)$. A plot of P versus λ is shown in Figure 4.3. Note that the probability of finding a linear relation is still significant at values of λ greater than one. For example, $P = .5$ when $\lambda = 2$. Furthermore, for similar values of λ, the probability of observing random classifications decreases much more rapidly for large values of n. This probability function

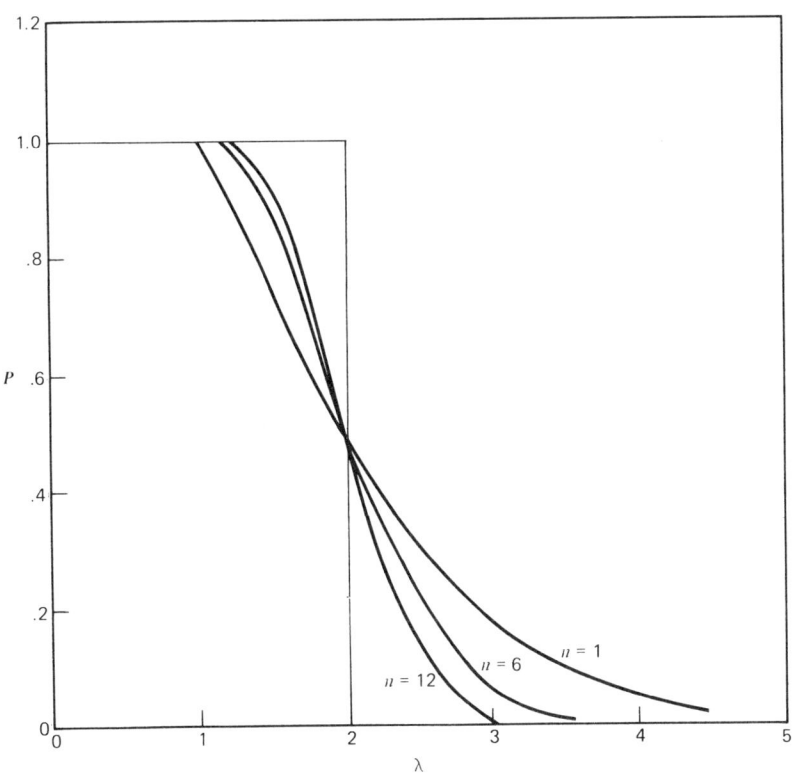

Figure 4.3 Probability of separating random, well-distributed data into classes plotted versus λ.

provides a gauge for determining the likelihood of observing a classifier that exhibits random classification. Any linear discriminant developed within this region would not be expected to indicate a relationship within the data.

To demonstrate the significance of random classifications, two data sets consisting entirely of random numbers were generated. Set 1 was generated using Gaussian distributed random numbers, while set 2 was generated using uniformly distributed random numbers. A total of 250 points was generated for each set. The sets were partitioned into training and prediction sets by randomly choosing half of the members for the training set and placing the remaining members in the prediction set.

Results for linear classifiers developed at various values of λ are shown in Table 4.1. It is evident from the table that the probability of obtaining a linear classification follows that predicted by equation 4.28. It is also interesting to note that the predictive ability for the members of the prediction set is no better than random.

Table 4.2 shows the manner in which P varies as a function of n and λ. Each column of the table shows how P changes for a constant value of λ as n increases. For the probability P to fall below 1% for $n = 15$, it is necessary that $\lambda > 3.0$. That is, N, the number of samples, must be at least $(15 + 1)(3.0) = 48$ for the probability of obtaining a separation between classes due to chance to fall below 1%. For data sets where λ is small, P remains large even for large values of n.

As a further demonstration we have constructed a data set consisting of

Table 4.1 Results of Developing Discriminants for Random Data[a]

Number in Training Set	Number of Cases	λ	Gaussian data				Uniform data		
			Theoretical P	Number Trained	P'	Predictive Ability	Number Trained	P'	Predictive Ability
35	20	1.67	.885	16	.80	51.2	15	.75	50.1
40	20	1.90	.625	11	.55	49.2	13	.65	51.5
45	30	2.14	.325	12	.40	48.2	6	.20	48.7
50	40	2.38	.126	2	.05	49.8	6	.15	47.8

[a] Data sets contained 250 members of 20 dimensions. Number of cases refers to the number of training sets used to measure P. Number trained refers to the number of sets which were separable. P' is the percentage of the total number of sets which were separable.

Table 4.2 Selected Values of P, the Probability of Separating Well-Distributed Data as a Function of n and λ[a]

n	P (2.25)	P (2.50)	P (2.75)	P (3.00)	P (3.25)	P (3.50)
3	.3633	.2539	.1719	.1130	.0730	.0461
5		.2120				.0207
6				.0577		
7	.3145	.1796	.0460		.0216	.0096
9		.1537		.0307		.0045
11	.2706	.1325	.0551		.0069	.0022
12				.0168		
13		.1147				.0010
15	.2498	.0998	.0330	.0093	.0023	.0005
17		.0871				.0002
18				.0052		
19	.2257	.0762	.0201		.0008	.0001
21		.0668		.0030		
23	.2051	.0587	.0124			
25		.0517				
27	.1871	.0456				

[a] Value in parentheses is λ.

300 samples, evenly divided between two classes and each sample represented by 40 variables. The data were generated such that a different number of variables define the relationship among them; these variables are called intrinsic variables. Removal of any one intrinsic variable destroys the relationship by essentially converting the data set to random numbers. If only the intrinsic variables are used to represent the samples, then the two classes are always separable with a linear discriminant; if any one of the variables is removed, then linear separability is lost. The method used to generate such a data set has been described previously (9).

Table 4.3 shows the results from linear discriminant functions, developed using a linear learning machine. The training and prediction sets were chosen randomly. Each entry in the table is the average of results obtained for training five sets with the indicated number of intrinsic variables, random variables, and λ values.

It is apparent from Table 4.3 that the predictive ability at any one value of λ is unaffected by changing the ratio of intrinsic and nonintrinsic variables as long as the total number of variables remains constant. For sets trained above a λ of 2.4, little if any effect on the predictive ability due to chance correlations

Table 4.3 Effects on Predictive Ability of Exchanging Random Variables and Intrinsic Variables[a]

Number of Variables			Predictive Ability						
			40	60	80	100	150	200	250
Total	Intrinsic	Random	(0.976)	(1.46)	(1.95)	(2.44)	(3.66)	(4.88)	(6.10)
40	40	0	47.4	49.4	53.5	58.2	72.8	80.4	83.2
40	30	10	49.0	50.3	53.5	56.8	73.6	81.2	82.8
40	20	20	48.5	48.2	52.8	59.2	76.0	85.0	85.2
40	10	30	50.6	51.0	53.9	59.3	69.6	80.2	80.8
40	5	35	51.9	52.6	58.0	61.7	73.3	81.0	85.6
Average predictive ability			49.6	50.3	54.3	59.0	73.9	81.6	83.5
Standard deviation			1.62	1.66	2.08	1.80	2.29	1.97	1.95

[a] The predictive ability reported is for the average of five independent trainings. Each column of predictive abilities is headed by the number of samples included in the training set and the resulting λ value.

would be expected. In these cases the predictive ability reflects the ability of the training set to represent the entire data set. The results in Table 4.4 indicate that this is the case. Shown are the results for developing discriminants using five randomly selected training sets. If the increase in the predictive ability for the classifiers was due to a decreasing probability of chance correlations, then it would be expected that the set with 200 members would lose its predictive ability less rapidly than the 100 member set. Similar decreases in predictive ability for each set would indicate that the differences in predictive ability between a set of 100 members and a set of 200 members arises because 200 members are more representative of the data than are 100. As the correlation between the decrease in predictive abilities is .98, the latter argument seems the most plausible.

Based on these observations, it might be assumed that the lower predictive ability for sets having a λ below 2.4 (Table 4.3) is due to these same effects. However, below 2.4 the predictive ability reflects the effects of chance correlations. To demonstrate this, the variance feature selection method (9) was used to select those variables that the classifier indicated as being intrinsic. None of the training sets with a $\lambda < 2.44$ were able to differentiate between the intrinsic variables and the nonintrinsic variables. In contrast, feature selection of the data above a λ of 2.44 showed that only the intrinsic variables were being used to classify the data. This demonstrates that the sets with low values of λ were developing relations in which the nonintrinsic variables played a

Table 4.4 Effects on Predictive Ability of Adding Uniformly Distributed Random Variables to Identical Pattern Vectors[a]

Number of Variables			Training Set of 100 Members		Training Set of 200 Members	
Total	Intrinsic	Random	λ	Predictive Ability	λ	Predictive Ability
20	20	0	4.76	76.6	9.52	94.2
25	20	5	3.85	77.0	7.24	91.6
30	20	10	3.23	68.2	6.46	88.6
35	20	15	2.78	63.4	5.56	86.2
40	20	20	2.44	59.2	4.88	85.0

[a] Each line reports the results for the average of five independent runs with five randomly selected training sets. The total data set contains 300 samples.

significant part. The low predictive abilities reflect the randomizing influence that the nonintrinsic variables imparted to the classifier.

The reliability of relations derived through use of pattern recognition techniques is dependent on proper application of the techniques. The basic arguments included here hold regardless of the form used for the discriminant function. These arguments concern the number of samples required to ensure that a nontrivial relationship is present.

It was demonstrated that although a relationship may indeed be present, classification attempts at low values of λ (< 3.0) fail to uncover it. The probability of obtaining a separating linear discriminant function for data in which no real information is present at low values of λ is high.

The inability to detect a relationship actually contained in the data was demonstrated through use of a feature selection technique that employs the results of the classification to determine those features that the classifier deemed most important. For the data sets with very small values of λ, the classifier indicated many of the random variables to be intrinsic to the classification process. This suggests that the feature selection technique, like the classification technique, is of little utility when the data are overdetermined. It is not possible to start out with an overdetermined data set and use the results of classification to lower the number of features to obtain an acceptable value of λ.

Remaining within the excess probability limits does not assure the reliability of the relationship; it only limits to acceptable levels the probability of developing a random relation. It is necessary to test the performance of any classifier using data that have not been employed in the development of the discriminant or the feature selection process. Only through successful classification of unknowns can the utility of a discriminant function be measured.

A final word of clarification is warranted. The assumption underlying these arguments is that the data are well distributed. If this is not the case, then the limitations are not as clearly defined. Such data may indeed be capable of supporting a linear relation at $\lambda < 3.0$. However, the degree to which these restrictions may be relaxed is not entirely clear. When a ratio under 3:1 is used, steps must be taken to demonstrate that a relationship truly exists. Failure to provide sufficient proof of performance is cause to doubt the validity of any results obtained from values significantly under 3.0. Clearly, nonparametric linear classifiers developed at λ values below 3.0 are unable to distinguish relationships contained within a set of well-distributed measurements. Results obtained from classifiers operated in this condition are of little utility unless other proof of performance is provided.

THE VARIANCE METHOD OF FEATURE SELECTION

Feature selection is defined as any method that reduces the dimensionality of a data space to form a feature space such that descriptors unnecessary for discrimination between the two classes are discarded, and those necessary for discrimination are retained. If the classes are linearly separable, then the feature space must be linearly separable after feature selection. In the case of a data set that is not linearly separable, the classification error function of the feature space must not be greater than that of the original data space.

The most well-defined methods of feature selection are statistical and assume prior knowledge of the probability density functions of the data classes. Diagonal, rotational, and other linear transformations are often employed in such methods.

In applications to structure–activity problems, very often no distributional assumptions can be made about the data. This necessitates the use of nonparametric methods of feature selection. Several nonparametric feature selection methods based on definitions intuitive rather than mathematical definitions have been reported (10–12). The variance method is also a nonparametric feature selection method. Unlike other methods it is based on a model of the process occurring when a linear learning machine develops a decision surface.

The variance method of feature selection identifies meaningful descriptors by ranking them according to the relative variation of the weight vector components within a series of trained weight vectors. The descriptors that correspond to larger relative variations can be identified and discarded in the order of their appearance in this list. The procedure employed to train a

The Variance Method of Feature Selection

series of weight vectors to be used in developing this list takes advantage of the properties of the linear learning machine.

A detailed discussion of the variance theory is given in the next section. The next few paragraphs discuss a method of applying the variance process.

Recall that data are represented as a set of vectors that have an extra dimension added to ensure a common origin. The first requirement in the development of the variance method is to optimize the value of this component. We refer to this component by the symbol x_n. During training, the x_n value is usually set to a convenient number that allows for both good predictive ability and a fast training rate. The variance method requires a specific range of x_n. The data set is trained using a value of x_n set equal to or less than the average of the absolute value for all the components in the data set. Successive weight vectors are then trained while the value of x_n is increased, and the results of the previous weight vector are used as the initialization for the next weight vector. The x_n increments typically range between 1 and 100. The value of the nth component of each weight vector is then plotted with respect to the value of x_n for the training trials. A typical plot is shown in Figure 4.4. The portion of the curve having the smallest slope is called the "permitted range" of x_n. The x_n component must fall into this range for successful implementation.

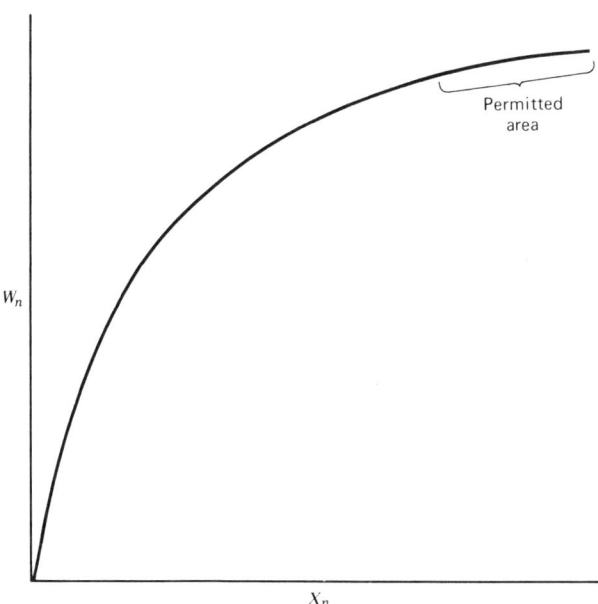

Figure 4.4 Example of a W_n versus X_n plot.

A second requirement is to train a sufficient number of different weight vectors while the x_n value is within the permitted range. The number of weight vectors to be developed depends on the data set. If large numbers of weight vectors are developed, fewer iterations of the algorithm are required. Generally, performance increases with a corresponding increase in the number of weight vectors used in the analysis.

Implementing the variance method by using linear learning machines has been described elsewhere (9). Application of the variance method to feature selection of drug data is described in Chapter 6. Therefore, only the results of applying the variance method to well-characterized artificially generated data sets is detailed here. These data offer an ideal model for demonstrating the effectiveness of the variance method; they are referred to as the DGEN data.

Five DGEN sets were generated such that each contained a total of 200 points in a 50-dimensional space. The only difference between the sets was the number of intrinsic dimensions. These dimensions define the separability of the data. The sets were generated containing 5, 15, 25, 35, and 45 intrinsic components. Each set was autoscaled such that the average value was zero and a standard deviation of 20 was obtained for each of the 50 components. Thus each data set is known to be linearly separable, and the number of intrinsic components, m, and nonintrinsic components, $50 - m$, are known in advance. The deletion of any of the intrinsic components during feature selection results in loss of ability by the classifier to separate the data, whereas elimination of any or all of the nonintrinsic components does not affect separability.

The performance of the variance method will be compared to that of the weight-sign feature selection method (10), an earlier nonparametric technique. Weight-sign feature selection was empirically developed as a result of the observation that in many cases the sign of the nonintrinsic weight vector components changed within a series of arbitrarily generated weight vectors.

In the variance method, a set of weight vectors are generated and the relative variation of each component is calculated using the following equations:

$$R_j = \frac{V_j}{\bar{w}_j} \qquad (4.29)$$

$$V_j^2 = \frac{1}{(n-1)} \sum_{k=1}^{n_k} (w_{jk} - \bar{w}_j)^2 \qquad (4.30)$$

Here j is the component index, k is the index for the weight vectors, $\bar{w}_j$ is the average value of the jth weight vector component, and n_k is the number of weight vectors trained.

The relative variations are then ordered from largest to smallest. Those components having the smallest relative variations are the most necessary for separability. Those components with the largest R_j values are the least necessary. Components are discarded in the order of decreasing R_j. The remaining components are those that give the classifier sufficient information to separate the classes in the data set. It may be necessary to develop and condense the ordered list of variations several times. This is usually an indication that an insufficient number of weight vectors were originally developed, or that the nth weight vector component is being allowed to vary excessively. For the best results the nth weight vector component should rank in the bottom 10 or 15% of the variance ranked list. The larger the value of x_n, the smaller the variance of the nth weight vector component.

The weight sign method relies on the fact that weight vectors obtained by the learning machine are a function of the order in which the data are presented to the classifier, the weight vector initialization, and the value of x_n. Varying one or more of these parameters leads to a number of different weight vectors, which are then subjected to weight–sign comparisons.

Six different initializations of weight vectors were used in the weight–sign feature selection trials: (1) $w_j = 1/\sqrt{n}$ for all j; (2) $w_j = -1/\sqrt{n}$ for all j; (3) $w_j = 1/\sqrt{n}$ for $j = 1, 2, \ldots, n - 1$, and $w_n = -1/\sqrt{n}$; (4) $w_j = -1/\sqrt{n}$ for $j = 1, 2, \ldots, n - 1$ and $w_n = 1/\sqrt{n}$; (5) $w_j = 0$ for $j = 1, 2, \ldots, n - 1$ and $w_n = 1$; (6) $w_j = 0$ for $j = 1, 2, \ldots, n - 1$ and $w_n = -1$. These are not particularly special; any six would have sufficed.

The weight–sign method was implemented by generating a weight vector using one of three initializations (1, 3 or 6) and the original order of the data set. A second weight vector was obtained by using the same initialization and a random scrambling of the order of the data set. The two weight vectors were subjected to weight–sign comparison, and opposite signed components were eliminated. The reduced data sets were repeatedly subjected to this scrambling and training procedure until no further eliminations were possible and each of the initializations had been used. The results are presented in the top section of Table 4.5. Only a fraction of the nonintrinsic components were eliminated by this procedure.

The variance feature selection method was then applied to the same five DGEN data sets with the results shown in the lower portion of Table 4.5. In each of the five cases, all the intrinsic components were identified and all the nonintrinsic components were discarded. The last two columns in the lower section of Table 4.5 refer to the number of times error correction feedback was used to derive a weight vector separating the classes; a substantial reduction was obtained when the nonintrinsic components were discarded. In all cases linear separability was shown by complete training to 100% recognition.

An example of the ordered list obtained from the variance method is shown in Table 4.6. This was obtained using only three weight vectors trained

Table 4.5 Results Obtained from Feature Selection Using Artificial Data Sets

	Data Set	Number of Steps	Descriptors Included	Initialization
Weight–sign method	1	4	35	1
		2	38	3
		4	38	6
	2	4	39	1
		1	44	3
		3	41	6
	3	2	31	1
		1	44	3
		4	40	6
	4	4	45	1
		2	46	3
		3	46	6
	5	1	49	1
		1	50	3
		1	50	6

				Number of Feedbacks	
				Initial	Final
Variance method	1	—	5	1024	76
	2	—	15	1722	356
	3	—	25	2332	804
	4	—	35	1990	1099
	5	—	45	2513	2138

on DGEN data set 1 using initializations 1, 3, and 5. This data set has five intrinsic components, numbered 1 through 5, and 45 nonintrinsic components, numbered 6 through 50. As can be seen from the table, the five dimensions responsible for separation appear lowest in the list while those that are not necessary for separation are all ranked higher.

It should be clear that the variance method offers an improvement over methods used previously. Further reports of results obtained using the variance method are given in Chapters 6 and 7. The next section contains a detailed description of the variance theory.

Table 4.6 Example Listing of Calculated Variations for Components of Artificially Generated Data

Descriptor Number	Calculated Variation	Descriptor Number	Calculated Variation
35	9.6258	7	0.1666
37	2.9912	25	0.1526
36	1.0876	41	0.1471
34	0.8762	38	0.1456
16	0.7122	20	0.1377
46	0.6876	31	0.1318
11	0.5851	43	0.1209
6	0.5024	40	0.1073
21	0.4493	45	0.1047
33	0.4034	28	0.1003
47	0.3927	27	0.0919
10	0.3924	51	0.0860
42	0.3752	18	0.0748
39	0.3673	22	0.0674
15	0.3555	29	0.0642
24	0.3037	26	0.0612
32	0.2748	8	0.0554
17	0.2689	30	0.0542
13	0.2226	9	0.0322
12	0.2209	49	0.0210
44	0.2190	4	0.0058
23	0.2043	2	0.0051
50	0.1904	3	0.0044
48	0.1824	1	0.0025
19	0.1822	5	0.0017

Derivation of the Variance Algorithm

Development of the variance method was in large part due to the ability to create a low dimensional model of processes occurring in the high dimensional spaces dealt within the application of pattern recognition to chemical problems. In this section the models that led to the development of the method are explained.

As an illustrative example, Figure 4.5 shows a data set with one intrinsic dimension, y, and one nonintrinsic dimension, x. The third dimension, z, is necessary to ensure a common origin. The two classes are linearly separable.

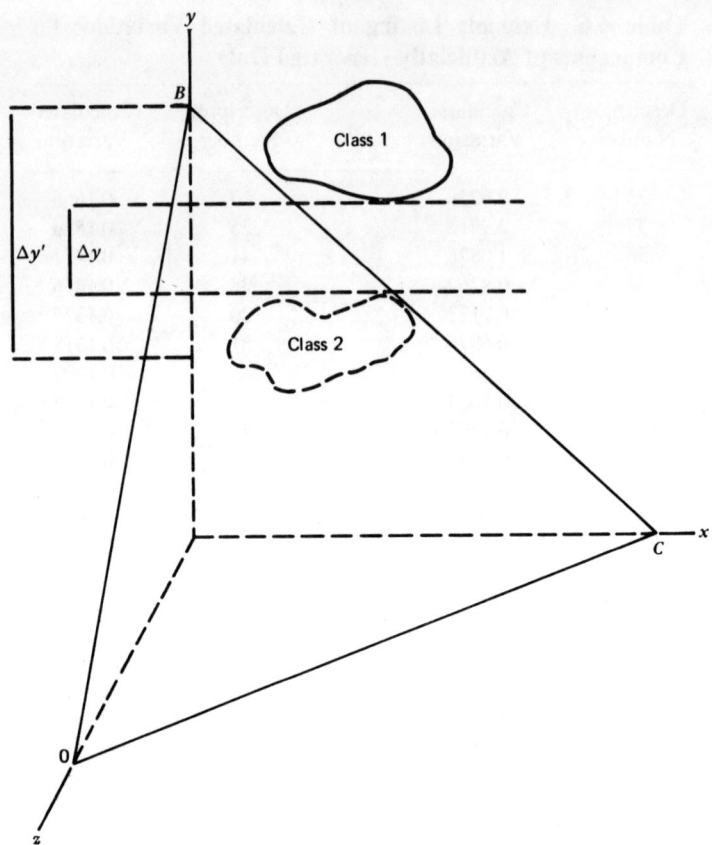

Figure 4.5 Orientation of the hyperplane separating a data set with one intrinsic and one nonintrinsic dimension.

The origin is point O. We let Δy be the minimum distance in the y dimension between the two data sets. Given no x component, then Δy represents the maximum range of the intersection of the separating surface, plane OBC, and the y axis. Note that the y dimension alone is sufficient to provide 100% recognition of the two groups. Note also that inclusion of an x component expands the range of y axis intersection to $\Delta y'$. As a result of this noise the surface now has an expanded range through which it may pass and still separate the two classes.

After addition of the x component the separating surface is constrained to (1) intersect the y axis only in the region labeled $\Delta y'$, (2) pass between the two classes without "touching" them, (3) pass through the origin at all times, and

The Variance Method of Feature Selection

(4) during any net movement retain the same "side" toward the respective classes. These constraints can be related to those imposed on a unit length vector (weight vector) that is perpendicular to the separating surface at the origin. It is this vector that is shifted during the training process. Since it must follow the movements of the surface to maintain its perpendicularity, as well as to satisfy the constraints, the plane can be considered to impose the above constraints on the weight vector.

The weight vector has three components, $\mathbf{W}_x$, $\mathbf{W}_y$, $\mathbf{W}_z$. It is to be demonstrated that when a number of decision surfaces, each separating classes 1 and 2, and each with an associated weight vector, are investigated, the relative variation in the magnitude of $\mathbf{W}_x$ (nonintrinsic component) is greater than the relative variation in the magnitude of $\mathbf{W}_y$ (intrinsic component). To attack this question it is necessary to understand the effect of changes in $\mathbf{W}_y$ as the data set is moved further from the origin.

Figure 4.6 represents the projection of the weight vector, $\mathbf{W}$, on the yz coordinate plane. We call the projection $\mathbf{W}_p$. Ψ is the angle between $\mathbf{W}_p$ and the z axis, θ is the angle between the projected separating surface, B, and the z axis and ϕ is the angle between B and $\mathbf{W}_p$. To classify the data correctly, the separating surface may pass only within the area $\Delta y'$. The absolute position $\Delta y'$ is fixed by the data set and $\mathbf{W}_p$ is constant for a given skew of the separating surface in the x dimension (the nonintrinsic dimension). An increase in the distance, z_1, causes a decrease in θ, an increase in Ψ, and, therefore, an increase in $\mathbf{W}_y$ and a decrease in $\mathbf{W}_z$. We define $\Delta\theta$ as the change in θ as the projection of the weight vector moves across the allowed range $\Delta y'$. $\mathbf{W}_y$ and $\mathbf{W}_z$

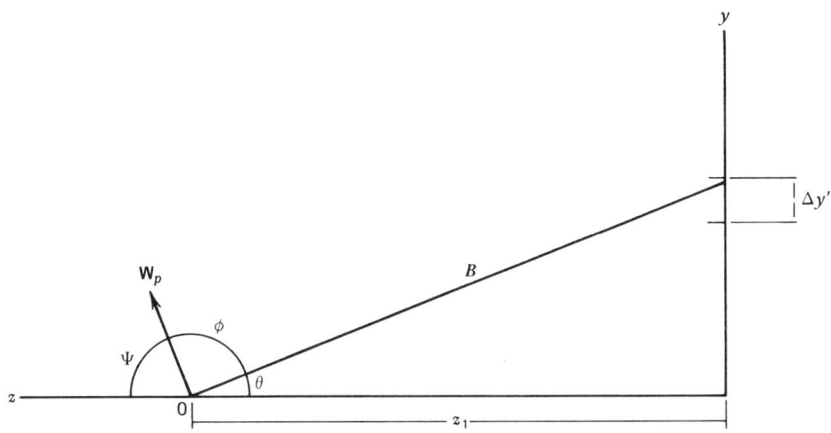

Figure 4.6 Diagram of the relation between Z_1 and $\mathbf{W}_p$ variability.

are defined analogously. Since $\Delta y'$ is constant, an increase in z_1 causes a decrease in $\Delta \theta$, a decrease in $\Delta \Psi$, and, therefore, a decrease in $\mathbf{W}_y$ and $\mathbf{W}_z$.

As z increases, the allowed variation in the $\mathbf{W}_z$ component for an allowable change in θ decreases. At large z_1 values the magnitude of $\mathbf{W}_y$ becomes independent of changes in θ. The only other contribution to $\mathbf{W}_y$ must then be due to changes involving movement of the weight vector in the xy coordinate plane. If the relative variation in $\mathbf{W}_x$ is greater than the corresponding relative variation in $\mathbf{W}_y$, then a means to distinguish intrinsic from nonintrinsic components is provided.

The variation of components $\mathbf{W}_x$ and $\mathbf{W}_y$ can be defined as V_x and V_y and the average magnitudes of these components are defined as $\overline{W}_x$ and $\overline{W}_y$. Then the relative variation of each component can be defined as follows:

$$R_x = \frac{V_x}{|\overline{W}_x|} \quad \text{where} \quad V_x^2 = \frac{1}{n-1} \sum_{k=1}^{n_k} (\mathbf{W}_{x,k} - \overline{W}_x)^2$$

$$R_y = \frac{V_y}{|\overline{W}_y|} \quad \text{where} \quad V_y^2 = \frac{1}{n-1} \sum_{k=1}^{n_k} (\mathbf{W}_{y,k} - \overline{W}_y)^2$$

where n_k is the number of weight vectors generated to approximate a spanning set. The definitions approach the definition of variance for a statistically large number of weight vectors. It is shown below that for a sufficient distribution of weight vectors in any range allowable by the geometry of the sets, $R_x > R_y$, that is, the nonintrinsic components vary more than the intrinsic components.

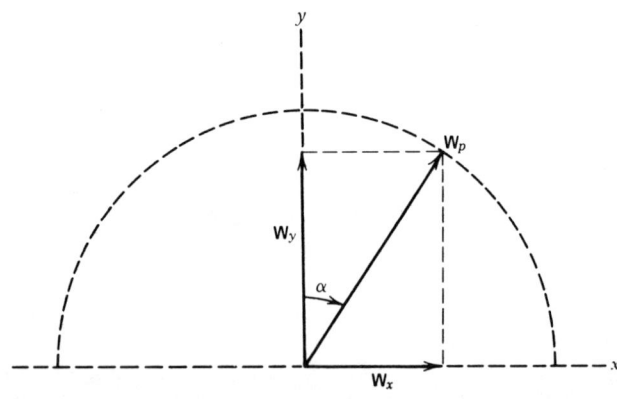

Figure 4.7 Projection of the weight vector on the xy coordinate plane.

The Variance Method of Feature Selection

In Figure 4.7 $\mathbf{W}_p$ is the projected weight vector from the n-dimensional pattern space (in this case three) into the $(n-1)$ space, the xy coordinate plane. Since the z component is fixed, this projection results in a vector of constant length whose x and y components are, respectively, $\mathbf{W}_x$ and $\mathbf{W}_y$. Only changes in the projected weight vector are of interest.

If we let the length of $\mathbf{W}_p = B$, then the vectors' components can be expressed as

$$\mathbf{W}_x = B \sin \alpha \tag{4.31}$$

$$\mathbf{W}_y = B \cos \alpha \tag{4.32}$$

where α is the angle between the projected vector $\mathbf{W}_p$ and the y axis (intrinsic dimension). We consider α to range between 0 and x. If α is continually varied over this range, then the average value of any component is given by its expectation value for that range. The expectation of a function $g(x)$ may be written as

$$g(x) = \int_{-\infty}^{\infty} g(x) P(x)/dx \tag{4.33}$$

where $P(x)$ is the probability of observing a specific value of x somewhere in the range of $g(x)$. For this problem

$$P(\alpha) = \frac{1}{\int_0^x d\alpha} = \frac{1}{X} \tag{4.34}$$

The expectation values are then

$$\langle \mathbf{W} \rangle_x = \int_0^x (B \sin \alpha) \frac{1}{r} d\alpha$$

$$= \frac{B}{X} [1 - \cos x] \tag{4.35}$$

$$\langle \mathbf{W} \rangle_y = \int_0^x (B \cos \alpha) \frac{1}{x} d\alpha$$

$$= \frac{B}{x} \sin x \tag{4.36}$$

The variance of components x and y over the same range is given by the second central product moment, which is

$$\sigma^2 = \int_{-\infty}^{\infty} g(x)^2 P(x) \, dx - \langle g(x) \rangle^2 \tag{4.37}$$

The x and y moments can be written as

$$\sigma_x^2 = \int_x^0 (B^2 \sin^2 \alpha)\frac{1}{x}\, d\alpha - \langle \mathbf{W}_x \rangle^2$$

$$= \frac{B^2}{x}\left[\frac{x}{2} - \tfrac{1}{4}\sin 2x - \frac{1}{x}(1 - \cos x)^2\right] \quad (4.38)$$

$$\sigma_y^2 = \int_0^x (B^2 \cos^2 \alpha)\frac{1}{x}\, dx - \langle \mathbf{W}_y \rangle^2$$

$$= \frac{B^2}{x}\left(\frac{x}{2} + \tfrac{1}{4}\sin 2x - \frac{1}{x}\sin^2 x\right) \quad (4.39)$$

The measurement of the moments defines the variance in the limit of an infinite number of uniquely generated vectors. If we let

$$\overline{W}_x = \langle \mathbf{W}_x \rangle; \qquad V_x^2 = \sigma_x^2$$
$$\overline{W}_y = \langle \mathbf{W}_y \rangle; \qquad V_y^2 = \sigma_y^2$$

then the relative variation as measured in the text, which is expressed as

$$R_x = \frac{V_x}{|\overline{W}_x|}; \qquad R_y = \frac{V_y}{|\overline{W}_y|} \quad (4.40)$$

can be seen to approach the relative standard deviation within the limit of an infinite number of uniquely generated vectors.

A graph of the ratio of R_y/R_x is presented in Figure 4.8. Note that within the ranges of $0° \leq \alpha < 90°$ and $270° < \alpha \leq 360°$ this ratio is less than 1. A value outside of these ranges violates one or more of the constraints imposed upon the plane.

If a sufficient number of vectors are generated within those ranges, such that a valid measurement of $V_{x,r}$ and $V_{y,r}$ can be made, then $R_x > R_y$. Also, for any number of generated vectors in the range $-45° \leq \alpha \leq 45°$ this is always true, since within this range $|\partial \mathbf{W}_x/\partial \alpha|$ is always greater than the corresponding change for $\mathbf{W}_y$. Within this range development of as few as three weight vectors may be sufficient to measure the variations.

In the range $-\pi/2 < \alpha < \pi/2$ the expected value for $\mathbf{W}_x$ is 0 while that for $\mathbf{W}_y$ is $2B/\pi$. If the allowable skewness due to the nonintrinsic component is symmetric about y, then the relative variation in the x component approaches infinity while the variation for the y component remains relatively small.

Additional nonintrinsic components show no net effect on these relations. This has been demonstrated using a four-dimensional data set, whose components were labeled as x_1, x_2, y, and z, corresponding to two nonintrinsic

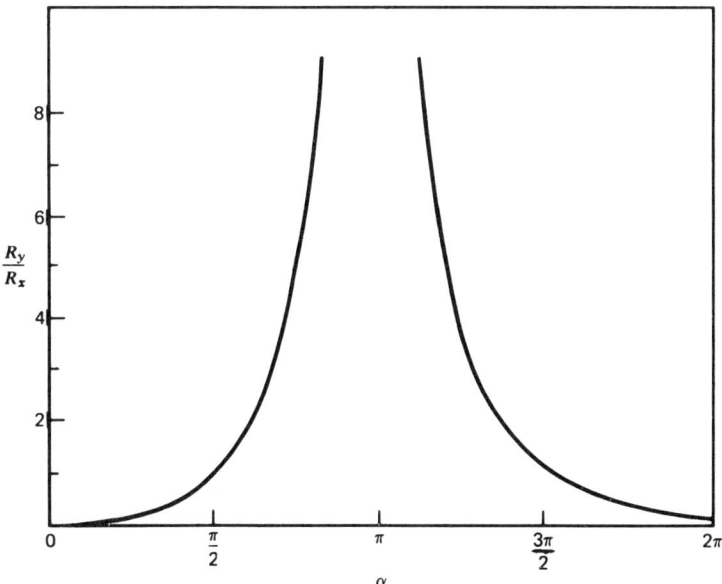

Figure 4.8 Plot of the ratio of variances versus α.

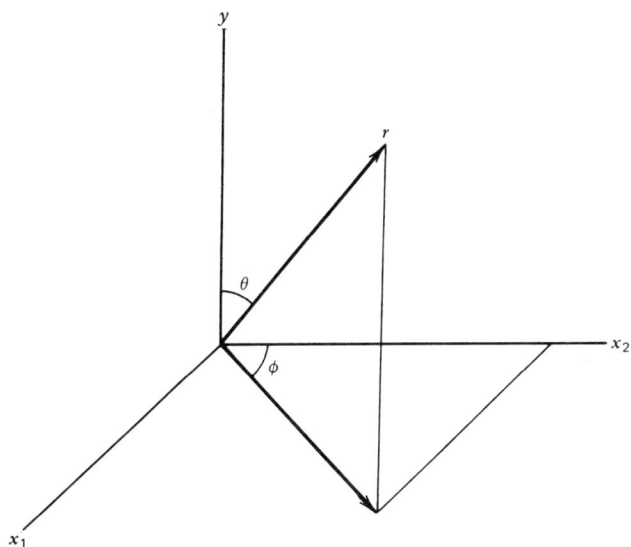

Figure 4.9 Coordinate system for a data set having one intrinsic and two nonintrinsic dimensions.

components, one intrinsic component and a component to ensure a common origin. Given that the data's z component was optimized as per the text, then Figure 4.9 shows the resultant projection along the z axis from $n = 4$ to $n - 1 = 3$ space. Since z is large its weight vector component is approximately fixed for any set of planes generated. The projected weight vector (r) is therefore of constant length. Using polar coordinates the relations of interest for $0 \leq \theta \leq x$ and any single value of ϕ are

$$\mathbf{W}_y = r \cos \theta \tag{4.41}$$

$$\mathbf{W}_x = r \cos \phi \sin \theta \tag{4.42}$$

$$P(\theta) = \frac{1}{\int_0^x d\theta} = \frac{1}{x} \tag{4.43}$$

$$\langle \mathbf{W}_y \rangle = \int_0^x [r \cos \phi \sin \theta] \frac{1}{x} d\theta$$

$$= \frac{r \cos \phi}{x} [1 - \cos x] \tag{4.44}$$

$$\langle \mathbf{W}_y \rangle = \int_0^x (r \cos \theta) \frac{1}{x} d\theta$$

$$= \frac{r}{x} \sin x \tag{4.45}$$

$$\sigma_x^2 = \int_0^x (r^2 \cos^2 \phi \sin^2 \theta) \frac{1}{x} d\theta - \langle \mathbf{W}_x \rangle^2$$

$$= \frac{r^2 \cos^2 \phi}{x} \left[\frac{x}{2} - \tfrac{1}{4} \sin 2x - \frac{1}{x} (1 - \cos x)^2 \right] \tag{4.46}$$

$$\sigma_y^2 = \int_0^x (r^2 \cos^2 \theta) \frac{1}{x} d\theta - \langle \mathbf{W}_y \rangle^2$$

$$= \frac{r^2}{x} \left[\frac{x}{2} + \tfrac{1}{4} \sin 2x - \frac{1}{x} \sin^2 x \right] \tag{4.47}$$

This set of equations differs from the two-space example by a constant involving ϕ. In the case where $\phi = 0$ the equations become those of the two-dimensional case. Note that in the expression for relative variation the ϕ dependence is lost and the equation becomes identical to that for the two-dimensional case. It would then follow that an n-dimensional coordinate frame could be defined such that the relations expressed in the previous two examples could be extended to a higher space. This would simply involve addition of terms to the constant. No such development is made since the two-dimensional model is adequate to indicate which parameters of the learning machine need to be optimized to select those features that are "intrinsic" to the separating process.

The significance of these arguments can be summarized as follows: (1) If the distance of the data set from the origin it constrains is increased, the change in the component of the weight vector corresponding to the intrinsic dimension is much less than changes in the position of the separating surface between the two classes, and (2) the variation in the intrinsic component of a weight vector reflects changes in the nonintrinsic component. The nonintrinsic component, however, is able to change more than the intrinsic. Therefore, measurement of the relative variation of the weight vector components at large values of z_1 should indicate those dimensions that offer no useful information concerning the separability of the sets.

REFERENCES

1. N. J. Nilsson, *Learning Machines*, McGraw-Hill, New York, 1965.
2. P. C. Jurs and T. L. Isenhour, *Chemical Application of Pattern Recognition*, Wiley-Interscience, New York, 1975.
3. Loftsgarden and Quesenberry, A Non-parametric Estimate of a Multivariate Density Function, *Annu. Math. Stat.*, **36**, 1049 (1965).
4. M. E. Hellmon, The Nearest Neighbor Classification Rule with a Reject Option, *IEEE Trans. Syst. Sci., Cybern., SSC-6*, 179 (1970).
5. T. M. Cover, Rates of Convergence of Nearest Neighbor Decision Procedures, *Proc. Annu. Hawaii Conf. Systems Theory, Isr*, 413, **1968**.
6. T. M. Cover and P. E. Hart, Nearest Neighbor Pattern Classification, *IEEE Trans. Inf. Theory, IT-13*, 21 (1967).
7. T. J. Wagner, Convergence of the Nearest Neighbor Rule, *IEEE Trans. Inf. Theory, IT-17*, 566 (1971).
8. P. E. Hart, The Condensed Nearest Neighbor Rule, *IEEE Trans. Inf. Theory, IT-14*, 515 (1968).
9. G. S. Zander, A. J. Stuper, and P. C. Jurs, Nonparametric Feature Selection in Pattern Recognition Applied to Chemical Problems, *Anal. Chem.*, **47**, 1085 (1975).
10. P. C. Jurs, Mass Spectral Feature Selection and Structural Correlations Using Computerized Learning Machines, *Anal. Chem.*, **42**, 1633 (1970).
11. P. R. Preuss and P. C. Jurs, Pattern Recognition Techniques Applied to the Interpretation of Infrared Spectra, *Anal. Chem.*, **46**, 520 (1974).
12. R. W. Liddell, III, and P. C. Jurs, Interpretation of Infrared Spectra Using Pattern Recognition Techniques, *Appl. Spectrosc.*, **27**, 371 (1973).

CHAPTER 5

A Software System to Implement Computer Assisted Structure–Activity Studies: The ADAPT System

The earlier chapters of this book discuss detailed descriptions of the various techniques used in computer assisted structure–activity studies. However, before a user can routinely apply these techniques, the procedures must be available in a convenient form. This chapter describes an interactive, modular computer system, called ADAPT, which was designed and implemented to provide convenient access to the capabilities necessary to perform computer assisted structure–activity studies.

Figure 5.1 shows the functional capabilities that a system must have to be useful for performing studies of structure–activity relations using pattern recognition techniques. As can be seen, a series of interrelated operations must be performed systematically using standard data structures that assure smooth flow between segments of the system. However, this simplified diagram avoids the question of the practicality of implementing these operations and the requirement of creating general algorithms that can be used on a wide variety of problems without modification.

To apply chemical structure information handling and pattern recognition techniques to studies of structure–activity correlations, a number of individual capabilities must be available to manipulate the data. The steps that are necessary are listed in order to show how they are interrelated.

1. Entry and storage of molecular structures. Entry is by sketching on the CRT screen of a graphics display terminal. Structures are stored on disc files as connection tables. Provision is made for the addition, deletion, alteration, and recall of structures. Substructures may also be entered and stored; these are used as input to substructure searching programs.
2. Entry and storage of a worklist of compound numbers for which molecular structure descriptors will be generated. A second list of compounds can also be entered independently and stored as the prediction list.

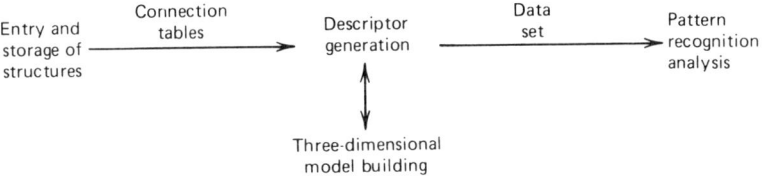

Figure 5.1 Computer system capabilities necessary to perform structure–activity studies.

3. Generation and storage of molecular structure descriptors from the connection table representations of the structures.
4. Generation of three-dimensional models of structures using a molecular mechanics approach.
5. Generation and storage of molecular structure descriptors from the geometrical representation of the structure.
6. Gathering together of a common set of descriptors for each compound in the working list to prepare a data set for analysis.
7. Analysis of the data set for correlation, separability, and so on using the following techniques drawn from statistics, nonparametric statistics, and pattern recognition:

   a. Multiple linear regression
   b. Bayesian discriminants
   c. Linear learning machines
   d. Nearest neighbor classification

8. Identification of which subset of the available descriptors are important enough to retain and which are unimportant enough to discard, that is, feature selection.

A general purpose system can be designed to implement all these steps. Obviously such a system is not restricted to studies that are directed at molecular structure–biological activity correlations, but has general utility.

If the data set to be studied consists of a set of molecular structures, then all the steps are necessary. If the data set to be studied consists of numerical measurements (e.g., mass spectra), then steps (3), (4), and (5) are unnecessary and the analysis begins with step (6). Whatever the source of the data, it will eventually be formed into a data matrix, where each row of the matrix contains all the measurements for one of the objects in the data set, and each column of the matrix consists of one particular measurement for all the objects. Preprocessing and prior feature selection of the data can be expressed as transformations of the data matrix, while classification and clustering are expressed as operations on the data matrix.

A potential pitfall in the implementation of pattern recognition techniques lies in the development of conventions concerning how these data matrices are constructed on input, labeled, stored, and accessed. If these operations are designed for a specific problem, then procedures suitable for one set of data may not suffice for others. Thus one of the primary prerequisites for a useful general purpose pattern recognition system is a general, data independent, file management system.

Figure 5.1 does not make clear the inherent diversity of the data handling problem. Not only must measurements from the transducer(s) be input, but they must be stored and labeled. Each data point must be given a class designation and identification number. Class designations must be easily assigned or modified. This ease of definition and redefinition is of the utmost importance in the overall data analysis. The source of the data is also important. Sources such as digitized spectra and complex molecular structures have widely different storage requirements. Since the operations performed on one type of data may bear little similarity to the operations performed on other types of data, a system designed with a high degree of modularity is required. Such a system is the ADAPT system. Each routine can execute independently, obtaining all necessary information either from a set of disc storage files or by interaction with the user. This mode of operation offers several advantages, the most obvious of which is a savings in memory.

The modularity provided through use of independent routines greatly decreases the complexity of the system and provides a means to incorporate additional algorithms into the system at any time. Thus the entire system is adapted to any user's individual requirements, since only those routines that are relevant to a particular problem need to be executed. In addition, these routines are relatively inexpensive to use because they do not require large scale facilities for execution. Finally, the routines are interactive in the sense that the user directs which manipulations are to be performed on the data.

ADAPT thus consists of a framework within which an unlimited number of routines can be supported. Each routine performs a specific, independent operation ranging from initial input of data to final output of results. The general utility of the system arises from the fact that the user has a large number of options to choose from, and he can conveniently interact with his data set.

Interaction with ADAPT is provided via a Tektronix 4012 CRT terminal. Data are stored in a series of defined files on cartridge discs. This allows fast access and ease of manipulation. The routines comprising the ADAPT system and the defined files accessed by those routines occupy approximately 2 million bytes of storage. The system is currently implemented using a 16-bit

MODCOMP II/25 computer located in the Department of Chemistry at The Pennsylvania State University.

The next few sections describe the individual routines within ADAPT that are used to perform operations such as data handling, class development, descriptor generation, data input, classification, and feature selection processes.

THE STRUCTURE FILE MANAGEMENT SYSTEM

The main data input to the ADAPT system is chemical structures. The routine SFILES (Figure 5.2) manages the input, storage, alteration, review, and deletion of molecular structures. Any of these tasks can be specified by the user during an interactive session with SFILES. Individual chemical structures are input to the system by sketching them in the familiar two-dimensional form on the CRT screen under control of the subroutine UDRAW. The information contained in the sketch is converted by UDRAW into a compressed connection table representing the atom types, bond types, and the bonding pattern. Rings are automatically perceived and labeled by a subroutine called FINDRG. A more detailed description of UDRAW's capabilities is given earlier in Chapter 3. The time required for the input of structures depends only on the user's proficiency in drawing structures on the CRT screen.

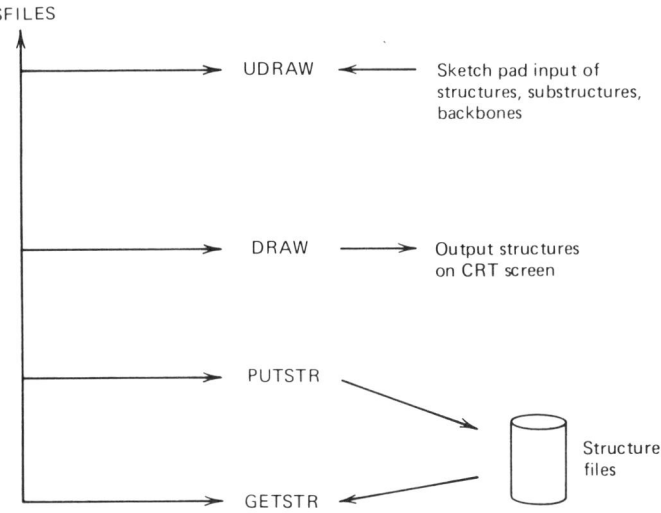

Figure 5.2 Entry, storage and review of structures using SFILES.

Once a structure is input through **UDRAW** to SFILES, it is given a direct access number (DAN number), and all information relevant to the structure is keyed to the structure's DAN number. After original entry to the defined files on disc, SFILES can be used to perform a number of tasks on any structure in the file. Any structure can be redrawn on the CRT screen in the same orientation in which it was originally input. If desired, the drawing can be labeled as to atom type, the numbering of the connection table, ring information, and so on. Thus there is usually no need to keep detailed drawings of the structures in notebooks, since they can be recalled and displayed at will. The stored information can be altered or deleted using SFILES commands. Unwanted structures are easily deleted from the structure files and new structures are easily added. SFILES is capable of maintaining a library of 1000 structures and associated auxiliary information as presently implemented. The size of the structure file is limited only by available mass storage.

To speed structural input, SFILES can store up to 10 different structural backbones, that is, structural forms that frequently occur in a series of molecules to be input. These can then be made to appear upon the initial **UDRAW** sketch pad and a complete molecule can be built up starting from this backbone. Thus a series of structurally similar compounds can be input without the need to redraw the base structure each time. In cases where the structures are to be modeled, the backbone fragments can be modeled just as if they were complete molecules. Modeling of the backbones prior to their use in the generation of larger molecules decreases the amount of time a molecular modeling program must spend on the larger molecule to place it in its lowest energy configuration.

As auxiliary information the user can assign a 20 character name and a 4 character label to a structure for his own reference, for example, index to his notebook files.

To minimize machine dependent features of SFILES, all disc input and output is contained within a pair of subroutines called GETSTR and PUTSTR. Thus transferral of SFILES to a new hardware system or different set of disc file conventions would require changing only these interfacing subroutines.

In addition to maintaining files of molecular structures, SFILES also maintains a library of up to 100 substructures. These files can be reviewed and entries can be added or deleted with a minimum of effort. The entries in the substructure file are input through UDRAW in the same way structures are input.

Once a structure file is created it can be queried by a series of tasks within SFILES that give various pieces of printed output regarding its contents. Information such as lists of the members and their respective DAN numbers can be generated, and lists of compounds having a given label can be selectively printed. The user can employ these routines to quickly obtain specific information concerning the data set saved on the disc. These queries, and all other commands, are entered via the CRT terminal.

CLASS DEVELOPMENT

Once the structural data are stored they must be organized into classes. ADAPT treats all problems as two-class problems. However, a routine is available that can quickly change class designations of the members. Therefore, multiclass problems are generally dealt with as a series of two-class problems.

ADAPT allows the user to create a list of 300 members for use in the subsequent analysis. This list is called the working list. This list consists of DAN numbers for the compounds contained on the structure files that have been chosen for inclusion in the active data set. The user generates the working list by supplying DAN numbers for the compounds, sorted into the two classes to be studied.

The class designations of the members of the working list can be changed after the working list has been constructed. Thus a series of studies could be performed using the same set of data, but with class memberships differently allocated, all without having to execute the class development program more than once. This is especially convenient for use in studies where the property being trained for is quantitatively or semiquantitatively known and the user wishes to train a series of discriminants with different threshold cutoffs between the yes and no classes.

Up to this point in the information flow through the ADAPT system the procedure is an interactive, real time process. The data set construction can easily be directed by the user and progress can be monitored using output from the CRT display or the line printer. Changes are extremely easy. The time required for a change is limited by the speed with which the user can specify the new parameters. Lists are constructed from commands input via the CRT or a card reader. Card reader input is especially convenient, since data sets can literally be constructed within seconds.

DESCRIPTOR GENERATION

The ADAPT system is designed to be modular so that descriptor generation routines are independent and new ones can be conveniently added. Two general classes of descriptors are developed—topological and geometric. Topological descriptors are generated from the connection table of a compound.

Geometrical descriptors are developed from a computer generated three-dimensional model. Each descriptor routine is independent and fetches the structural information it needs from the structure files and outputs the descriptor to a set of descriptor files. All descriptors are stored in a main descriptor file. Each is labeled with a name and a numerical flag for identification. The name is a four character alphabetic label provided by the descriptor generation routine, and the flag is a numeric value assigned by the descriptor generation routine. Together they allow for complete identification of the contents of a descriptor file.

Descriptors are stored as real or integer values, according to which is natural for the descriptors being generated. Each descriptor routine can work on one compound at a time, with full intermediate printout if desired, or on all the compounds forming the working list. If descriptors are developed for the entire working list, the user retains complete control over which, if any, of the descriptors will be stored on disc, and, if so, where.

A utility routine, DFILES, is a general file maintainence and interrogation routine for the descriptor files. It is used to initialize the descriptor files, to review the contents on any descriptor file, to produce a list of all the descriptors currently stored, to delete a descriptor from storage, to generate sets of test descriptors and store them for testing purposes, to compress the main descriptor file, and for other maintenance operations.

The principles embodied in the descriptor development routines are described in detail in Chapter 3. The following paragraphs give the name and a brief description of each routine currently implemented.

DMFRAG: Calculation of Molecular Fragments

This routine generates the following 16 molecular fragments for each compound:

1. The number of atoms in the structure.
2. The number of carbon atoms in the structure.
3. The number of oxygen atoms in the structure.
4. The number of nitrogen atoms in the structure.
5. The number of sulfur atoms in the structure.
6. The number of fluorine atoms in the structure.
7. The number of chlorine atoms in the structure.
8. The number of bromine atoms in the structure.
9. The number of iodine atoms in the structure.
10. The number of phosphorus atoms in the structure.
11. The number of bonds in the structure.
12. The number of single bonds in the structure.
13. The number of double bonds in the structure.

Descriptor Generation

14. The number of triple bonds in the structure.
15. The number of aromatic bonds in the structure.
16. The molecular weight of the compound.

The number of occurrences of each fragment in each compound in the work list is calculated and the operator can then store any of these fragment descriptors on the data files. The labels attached to the fragment descriptors are "FRAG" and the flag values are the numbers listed above.

DMSSS: Molecular Substructure Searching

This routine generates descriptors using a substructural search algorithm. The substructure(s) to be used are entered into the substructure storage area using SFILES. Their DAN numbers are supplied to DMSSS during execution in response to a query. The substructure searches can be done taking ring information into account or not, as the user wishes. (MFLAG = 0 for ring information being considered, MFLAG = -1 if not considered.) The molecular connectivity of the substructure as imbedded in the structure can be calculated as the descriptor if desired. (MC = 0 if molecular connectivity is not to be calculated, MC = 1 if it is to be calculated.) Either all occurrences or only the unique occurrences of the substructures can be found.

For storage, the user supplies a list of library access numbers into which the descriptors are to be put. The labels attached to the descriptors are "SSS" when the number of occurrences comprises the descriptor, or "ENVR" when the molecular connectivity of the substructure comprises the descriptor. The flag values are the DAN numbers of the substructure used to develop the descriptor, with a plus sign for MFLAG = 0 and a minus sign for MFLAG = -1.

DMCON: Molecular Connectivity

The DMCON routine generates molecular connectivity (M.C.) descriptors for the supplied compounds. It calculates six variants:

1. Path 1 M.C. for all bonds in the structure.
2. Path 1 M.C. corrected for rings.
3. Path 1 M.C. calculated using the valences of heteroatoms and corrected for rings.
4. Path 2 M.C.
5. Path 3 M.C.
6. Path 4 M.C.

The labels attached to these descriptors are "MOLC" and the flag values are the index numbers above.

DMVOL: Molecular Volume

This routine calculates the molecular volume of the supplied compounds by summing the volumes of spheres of radii equal to the van der Waals radius for each atom in the structure. Overlaps are subtracted from the sum. The calculation can be done using modeled coordinates for each compound or using standard bond lengths, at the user's option. The label attached to the generated descriptors is "MOLV" and the flag value is set to 1.

DMGEO: Molecular Geometry

This routine calculates the principal moments of each component supplied. They are arranged in decreasing order; the largest is called X, the intermediate value is called Y, and the smallest is called Z. The three ratios X/Y, X/Z, and Y/Z are also calculated. The label attached to the six descriptors is "GEOM" and the flag values are 1 through 6. The geometrical descriptors are calculated from the strain energy minimized model developed by the molecular modeling routine described in Chapter 3.

Through the application of the descriptor development routines, files of descriptors are built up for each molecule contained in the working list. These lists of descriptors, or subsets of them, can then be used for pattern recognition analysis by the remainder of ADAPT.

In addition to structural input, ADAPT also can accept information in a generalized vector format. This mode of input is implemented to mimic the descriptor development routines. Since each descriptor generation routine uses a common storage area, data external to the ADAPT descriptor development routines may be treated as data from another descriptor package.

The external descriptor input routine, DEXTR, allows the input of descriptors from outside the ADAPT system. It also implements a number of utility tasks. Descriptors can be entered as an independent set of data to be used alone, for example, for testing the pattern recognition analysis routines. Alternatively, the descriptors can be entered to correlate with the worklist. They can be entered either by compound or by descriptor, using formatted or free input, at the user's option. They can be real or integer and are stored with the label "EXTR".

These external descriptors could be from digitized spectra, physical measurements, data from on-line experiments, or descriptions of molecules in the working list that were generated by means other than the ADAPT descriptor development routines. For example, if a physiochemical measurement was determined experimentally for each of the molecules in the working list, it could be entered into ADAPT through DEXTR, and used in the analysis along with the internally generated descriptors. In this manner a basic data set containing information from a wide variety of sources can be constructed.

Two utility routines allow the user to investigate descriptors individually or to manipulate them arithmetically. Routine MATH allows previously generated descriptors to be altered one at a time by exponentiation, the taking of logarithms, or the taking of the square root. Two descriptors may be added, subtracted, multiplied, or divided as well. Routine CORCOF calculates the linear regression cofficients between all pairs of stored descriptors, printing out the results. A combination of MATH and CORCOF can be used in a flexible and convenient way to investigate a data set of relations between pairs (or triples, etc.) of descriptors.

At this point a data set consists of a working list and a number of files of descriptors saved independently on disc files. To proceed with the analysis of the data, a data matrix must be formed.

ACTIVE DATA SET: FORMATION, FEATURE SELECTION, AND CLASSIFICATION

To this point all operations have been those necessary to create or manipulate descriptors. From this point on we are concerned with the actual manipulations to be performed using these descriptors. The first step in this process is to form an active data set from those measurements that the user has determined to be significant. In general not all the measurements made on a system are of interest. The user may wish to utilize only certain pieces of this information in the study. The routine COLATE allows the user to pick any subset of the available descriptors and form an active data set. Thus the user can choose to include in the active data set only those descriptors that appear to be most important. Alternatively, the user can test several different subsets of descriptors to find the set that performs best. Once created the active data set can then be treated using a whole spectrum of pattern recognition techniques.

COLATE stores the active data by descriptor. Each component of the descriptor contains the value for a given element. These components follow the order that appears in the working list (generated by CLSMKR, or DEXTR). As an example of this type of storage, we consider the case where the user wishes to store 300 spectra each having 100 peak positions. COLATE would store these as a 100 by 300 matrix where the rows correspond to the intensities at peak positions labeled 1 to 100. The columns, numbered 1 to 300, correspond to the spectra from which the peaks were taken. This storage format is easily accessed by a classifier or any other program through the use of two arrays containing the indirect addressing pointers to the active rows and columns. The active column array is a list of which columns (molecules, spectra, etc.) of the data matrix are to be used. The active row array contains a list of the descriptors to be used. Initially, these arrays are set such that all the rows and columns defined by the working list and by COLATE are active. The user may then deactivate any of these based on results of feature selection or classification.

Service routines exist that quickly perform such operations as generation of training and prediction sets, changing of class designations, and removal of unwanted descriptors. Many of these operations simply involve redefinition of the active rows and columns of the data set. In this manner, manipulation of the active data set is quickly and easily accomplished.

After it has been collated, the data matrix is ready for analysis. ADAPT supports a wide variety of preprocessing or prior feature selection methods that can be used to transform the data. COLATE can be used to implement the results from prior feature selection methods by allowing the user to specify which descriptors are to be used. There are a number of routines available that calculate correlation coefficients, Fisher ratios, separation abilities of individual descriptors, separation abilities of all descriptors taken pairwise, a statistically based U statistic quantity, and others.

After it is preprocessed, the data set is presented to a discriminant development or other classification routine. ADAPT can support any of the common methods: linear learning machine, K nearest neighbor classifier, least squares procedures, clustering algorithms, and so on.

The weight vectors from the discriminant development routines can be used in a feedback loop to perform variance feature selection. This allows identification of a minimum set of descriptors that support linear separability for the data set. Alternatively, feature selection can be performed based on the user's own review of the results from any of the discriminant functions or classifiers. The important point is that the user has a wide number of choices and can conveniently exercise ingenuity to extract meaning from large sets of high dimensional data.

SUMMARY

This chapter describes the system called ADAPT, which was designed to overcome the two major problems which arise in the application of pattern recognition to chemistry. The first problem is data management. One cannot interact with the data unless the tools are available to efficiently mold it into useful forms. The use of the defined file system with a fixed data structure is one solution to this problem. The old files are easily expanded and new files are easily added as the system grows.

Summary

The second problem is that of having a system that restricts the user to a fixed set of preselected procedures. This is solved by using a set of modular, independent routines that interact with a file management system. Expansion of the capabilities of the system is relatively easy because of the modular construction. Newly developed routines can access the data easily, since it is already transformed into a useful form by the file maintenance routines. Use of independent routines also allows the system to be executed on a relatively small laboratory computer. This makes the system cheaper to operate and usually improves the access of the computer to the user, thus enhancing interactive computing. It is felt that with the development of this type of computer system, a major barrier to studies of chemical applications of pattern recognition techniques has been removed. Thus the major questions of the ultimate utility of these techniques can be addressed.

CHAPTER 6

Drug Structure–Activity Relation Studies

The assumptions that underlie the pattern recognition approach to structure–activity studies are similar to those of Hansch analysis. The factors governing the activity of a compound are viewed as combinations of a molecule's electronic, steric, and lipophilic properties. Compounds that have similar combinations of these parameters have similar activities. The standard parameters used to represent these properties have been various linear free energy related constants, such as the lipophilicity index, Taft constants, and Hammett constants.

Most pattern recognition analyses do not employ the linear free energy related parameters because of the difficulty in determining them for the large number of compounds used in most analyses. The majority of pattern recognition studies have based their approximation of biological similarity on parameters that can be derived directly from the structure of the molecule. Examples of these parameters were given in Chapter 3.

This chapter and Chapter 7 give examples of the types of studies that can be performed using pattern recognition techniques. Several other studies of this type have been performed in the last few years. Hansch et al. (1) have published a review discussing the application of hierarchial clustering techniques to the selection of substituent constants. Ting et al. (2) have reported correlations between the low resolution mass spectra of 66 drugs and their activity as sedatives or tranquilizers. Kowalski and Bender (3) and Chu et al. (4) have reported applications of pattern recognition to investigations of structure–activity relations by using substructural parameters as descriptors of biological action. Cammarata and Menon (5,6) have applied several methods of analysis to a set of compounds of accepted therapeutic utility and have discussed the application of pattern recognition methods to structure–activity studies of pharmaceuticals. Other examples of using structurally derived parameters in studies involving pattern recognition have been published (7,8).

APPLICATION TO PSYCHOTROPIC AGENTS

The studies discussed below give examples of the type of information provided by pattern recognition analysis.

APPLICATION TO PSYCHOTROPIC AGENTS

The present study is concerned with the implementation of adaptive binary pattern classifiers to distinguish between drug molecules that exhibit activity as sedatives and those that exhibit activity as tranquilizers. The classifier bases its decision entirely on information available from a standard two-dimensional structural representation of the molecule. In the present implementation no geometrical descriptors or physical property descriptors other than molecular weight are employed, although their use is not precluded by the techniques being used. The results of classifications are then further used to deduce which of the given parameters are most effective in the determination of a given activity.

The Data Set

The set of drugs used in this study consisted of 219 compounds selected from a standard reference (9). The set contained the 140 tranquilizers and 79 sedatives given in Table 6.1. A number of different parent ring types are represented, including phenothiazines, indoles, benzodiazepines, barbiturates, heterocyclic butyrophenones, nitrogen heterocycles, non-nitrogen heterocycles, and diphenylmethane derivatives. All the compound types represented and the number of each type are given in Table 6.2.

Many medicinal chemists disagree on the precise classification of a large segment of the psychotropic agents and, therefore, the classifications of sedative and tranquilizer are not exacting. Many drugs show activities in both of these classes, as well as in others, for example, hypotensive and muscle relaxant. Generally tranquilizers are classified as being either major or minor tranquilizers, while many sedatives show hypnotic action. The method used in this paper to classify the compounds' major action as sedative or tranquilizer is based on the information in reference 9. In this reference compounds were classified as major tranquilizers (TMa), minor tranquilizers (TMi), tranquilizers (T), sedatives (Sed), hypnotics (Hyp), or sedative-hypnotics (Hyp-Sed). The classification rules were applied as follows: (1) If the compound was TMa, TMi, or T, then classify as tranquilizer; (2) if the compound was Hyp, Sed, or Hyp-Sed, then classify as sedative; (3) if the compound was a combination of activities, such as (T Sed), (T Hyp), (TMi Sed), classify as tranquilizer; (4) if in any of the multiple classifications given there

Table 6.1 Compounds in the Data Set

Tranquilizers

1 A124	2 Acepromazine
3 Aceprometazine	4 Acetophenazine
5 Butaperazine	6 Butyrylpromazine
7 Carphenazine	8 CB 1519
9 CB 1658	10 Chlorimipiphenine
11 Chlorproethazine	12 Chlorpromazine
13 Chlorpromazine[a]	14 Ciba 17040
15 CPO 12	16 Cyamepromazine
17 Cyclophenazine	18 Dichlorpromazine
19 Dixyrazine	20 Ethylisobutrazine[b]
21 Fluorophenothiazine	22 Fluphenazine[b]
23 Fluphenazine	24 Fluphenazine[c]
25 Hepltylpromazine	26 Homophenazine
27 KS-33	28 MD 5501
29 Mepazine	30 Mesoridazine
31 Methiomeprazine	32 Methophenazine
33 Methotrimeprazin	34 Methoxypromazine
35 Oxaflumazine	36 P 824
37 P 1030	38 Perazine
39 Perimetazine	40 Perphenazine
41 Perphenazine[a]	42 Phenazin
43 Phenazine	44 Pipamazine
45 Piperactazine	46 Piperidochlor-[d]
47 Prochlorperazine	48 Prometazine
49 Promazine	50 Propiomazine
51 Propiopromazine	52 Ridazine
53 R.P. 3300	54 R.P. 4627
55 R.P. 6696	56 R.P. 9153
57 SA 124	58 SAF 5657
59 SKF 6333	60 T 412
61 Spiclomazine	62 Thiethylperazine
63 Thiopropazate	64 Thioproperizine
65 Thioridazide	66 TPN 12
67 Trifluoperazine	68 Trifluoperazine
69 Triflupromazine	70 Triflutrimeprazine
71 Trimeprazine	72 Valeroyl-perazine
73 Win 13,645-5	74 Chlorproheptadiene
75 Clomacran	76 Clopenthixol
77 Clothiapine	78 Clothixamide
79 Cyanothepin	80 Desmethyl-doxepine
81 Doxepin	82 Flupenthixol
83 G 22150	84 ID 22

Table 6.1 (*Continued*)

Tranquilizers	
85 Luxapine	86 Trifluthepin
87 Trimepramine	88 Xanthiol
89 Bishomoreserpine	90 Reserpedine
91 Methyl-18-ketore	92 Raujemidine
93 Raunescine	94 Renoxidine
95 Rescinnamine	96 SU5171
97 8842	98 SU10704
99 Raubasine	100 Renanserin
101 Benzindopyrine	102 DIM
103 3-IAAR	104 IN 399
105 Milipertine	106 Oxypertine
107 PI 11	108 Solpyertine
109 Dimechrom	110 Bromazepam
111 Chlorazepate	112 Chlordiazepoxide
113 Cloazepam	114 Cloxazolazepam
115 CT 5104	116 Cyprazepam
117 Diazepam	118 Isoquinazepon
119 Lorazepam	120 Medazepam
121 Nitrazepam	122 Nitrazepate
123 Oxazepam	124 Oxazolam
125 Prazepam	126 RO5-2180
127 RO-53027	128 Sulazepam
129 Temazepam	130 Tetrazepam
131 Aceperone	132 AHR 1900
133 FR-33	134 Diphenchloxazine
135 Misaflur	136 Prothipendyl
137 Trioxazine	138 Captodiame
139 Phenyltoloxamine	140 Cintrlamide

Sedatives	
1 Profenamine	2 Promethazine
3 Cloxypendyl	4 Fenoharman
5 Cannabioerol	6 D-58S1
7 Lorazepam	8 Allobarbital
9 Alphenal	10 Amobarbital
11 Aprobarbital	12 Barbital
13 Butalbitol	14 Butethal
15 Butallylonal	16 Cyclobarbital
17 Cyclopal	18 Febarbamate
19 Heptabarbital	20 Hexethal

(*Continued*)

Table 6.1 (*Continued*)

Sedatives	
21 Hexobarbital	22 Mephobarbital
23 Methabarbital	24 Methitural
25 Methohexital	26 Nealbarbitone
27 Pentobarbital	28 Phenobarbital
29 Probarbital	30 Secobarbital
31 Talbutal	32 Thiamylal
33 Thiopental	34 Vasalgin
35 NSD 2023	36 Anileridine
37 Catapresan	38 CHI 21
39 CHI 34	40 CHI 38
41 CHI 42	42 Clomethiazole
43 Dichlormethymone	44 ES 708
45 Ethinazone	46 Glutethimide
47 Homochlorcyclizine	48 K-2004
49 LB 50160	50 Mecloqualone
51 Methaqualone	52 Methyprylone
53 Oxypenayl	54 Tetridin
55 Thalidomide	56 Ethomoxane
57 Paraldehyde	58 WB 4123
59 Tricetamide	60 Caitamine
61 RD 6020	62 CD 6030
63 Chloral hydrate	64 Dispranol
65 Ethinamate	66 Mebutamate
67 Meprobamate	68 Nisubamate
69 Ethchlorvynol	70 Methylpentynol
71 Petrichloral	72 Acetylcarboromal
73 AEC	74 Carbromal
75 Bromisovaium	76 Ectylurea
77 IPC	78 Valnoctamide
79 Chlorethate	

[a] Sulfoxide.
[b] Decanoate.
[c] Enanthate.
[d] -Promazine.

is a preponderance of one class over the other, the compound is classified as belonging to the major class.

In no case was the classification changed so that a more favorable result with respect to recognition rate was effected. The ability of the pattern recognition approach to deal with such a heterogeneous data set may be one of the strengths of the technique.

Table 6.2 Structural Classes Represented in the Data Set

Compound Type	Number of Tranquilizers	Number of Sedatives
Phenothiazines	73	2
Phenothiazine analogues and isomers	15	1
Indoles		
Reserpine and derivatives	10	0
Haramine and derivatives	1	1
Others	9	0
Cannabis derivatives	0	1
Other heterocycles		
Chrome derivatives	1	0
Benzodiazepines	21	2
Barbiturates	0	27
Heterocyclic butyrophenones	3	1
Other N heterocycles	4	20
Benzodioxone derivatives	0	1
Non-N heterocycles	0	2
Aromatic compounds		
Diphenyl methane derivatives	2	0
Benzoic acid derivatives	0	1
Others	1	3
Aliphatic compounds		
Glycols	0	2
Carbamates	0	4
Carbinols	0	3
Amides and hydrazines	1	7
Others	0	1
	140	79

The success of the application of binary pattern classifiers to structure–activity correlations depends on the method used in describing the molecular structures. In this study three types of descriptors were employed: binary and numeric fragment descriptors, binary substructure descriptors, and topological descriptors. The 69 descriptors are given in Table 6.3. Each descriptor is contained in a minimum of 10% of the drug structures, and in no case does any one descriptor contain enough information to successfully classify the compounds.

While the nature of most of the descriptors is evident, some of them require further explanation. Descriptor 15, total weighted bond length, is calculated

Table 6.3 Descriptor List[a]

1. Molecular weight
2. Number of nonring carbon
3. Number of nonring oxygen
4. Number of nonring nitrogen
5. Number of nonring sulfur
6. Number of fluorine
7. Number of chlorine
8. Number of oxygen
9. Number of nitrogen
10. Number of sulfur
11. Number of carbon
12. Number of C=C
13. Number of C—C
14. Number of phenyl bonds
15. Total weighted bond length
16. —OC(=O)—
17. —C(=O)—
18. Aromatic ring 〉C—
19. 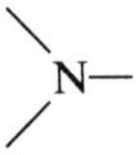 〉N—
20. 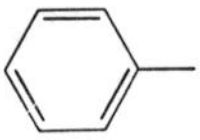 ⌬—
21. 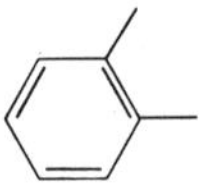 ⌬(CH₃)— (ortho methylphenyl)
22. 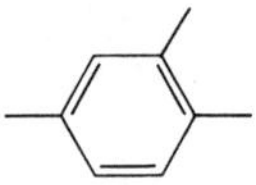 —⌬— (disubstituted methylbenzene)
23.  —N⌒N— (piperazine)
24. Ring 〉S
25. Ring 〉N

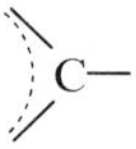

26. Ring 〉NCH₂—
27. Ring 〉O
28. 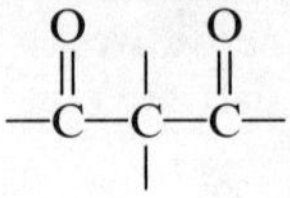 —C(=O)—C—C(=O)—
29. 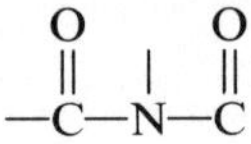 —C(=O)—N—C(=O)—
30. 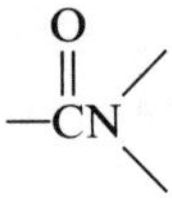 —CN〉(=O)
31. 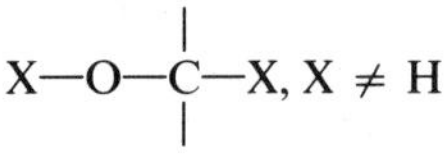 X—O—C—X, X ≠ H
32. 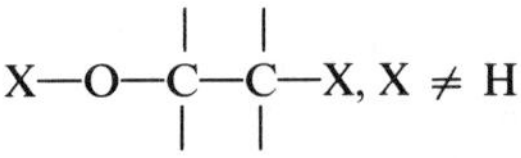 X—O—C—C—X, X ≠ H
33. 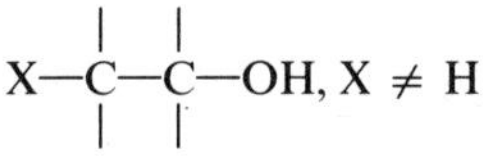 X—C—C—OH, X ≠ H
34. —OH
35. 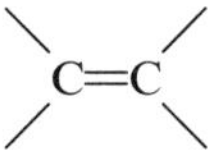 〉C=C〈
36. 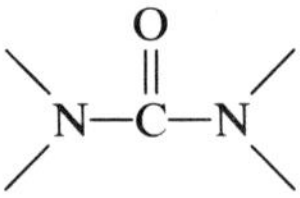 〉N—C(=O)—N〈
37. 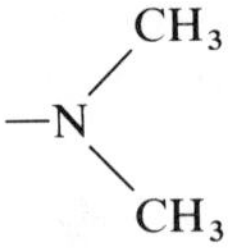 —N(CH₃)(CH₃)
38. 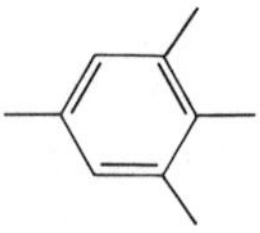 ⌬ (trimethylbenzene)

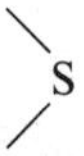

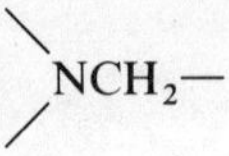

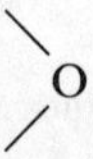

144

Table 6.3 (*Continued*)

39. Ring $\diagup\!\!\!\!\!^{\diagdown}\!\!\text{N}\!-\!\text{CH}_3$
40. —C—C—C— [b]
41. —C—C—C—
42. —C—C— [c]
43. —C—C—
44.  ⟨○⟩N any substitution

45. Longest chain of nonaromatic carbon
46. Terminal X—CH$_3$ where X ≠ a carbon chain
47. Y—C—C—X, X ≠ H; —C— Y = H or c

48. Aromatic ring $\diagup\!\!\!\!\!^{\diagdown}$C—
50. NH$_2$—
52. —C—
54. $\diagup\!\!\!\!\!^{\diagdown}$N—
56. $\diagup\!\!\!\!\!^{\diagdown}$CH—

Topological Descriptors [d]

58. CH$_3$—
60. —CH$_2$—
62. —NH—
64. —O—
66. Aromatic ring $\diagup\!\!\!\!\!^{\diagdown}$CH
68. —C(=O)—

---

[a] Descriptor numbers 1–15, 45, and 46 are numeric descriptors; the others through 47 are binary. [b] Value is 1 if CH$_3$CC is present, 2 if —CCC— is present, and 3 if both are present. [c] Value is 1 if CH$_3$C is present, 2 if —CC— is present, and 3 if both are present. [d] The topological descriptors were developed in pairs—bond and weighted—for the fragments shown.

by summing 4 for each single bond, 3 for each phenyl bond, 2 for each double bond, and 1 for each triple bond in the molecule, and dividing by 2. Descriptor 18 is a binary descriptor that indicates whether an aromatic ring is branched. Descriptors 20, 21, 22, and 38 are binary descriptors that indicate whether or not these explicitly described substructures are present. Descriptor 48 differs from 18 in that 48 is a topological descriptor. Descriptor 41 differs from 40 and 43 from 42 in that 41 and 43 are binary substructure descriptors and the others are coded as shown and are numeric descriptors.

Thus the raw data set consists of 219 drug structures each coded with 69 descriptors. Preprocessing of the raw data prior to training consisted of normalizing, autoscaling, and variance weighting. Normalizing consisted of multiplying each component of the data set by a factor such that the average value of all nonzero components was equal to 20. Then each descriptor was truncated to an integer value. This process yielded a normalized integer valued data set called the NDATA set. Secondly, the normalized data were subjected to autoscaling and variance weighting, thus giving a normalized, autoscaled, variance weighted data set called the NAVDATA set. After the autoscaling the normalized data were multiplied by a factor of 20 and truncated. The normalized and autoscaled data were variance weighted and multiplied by a scaling factor of 500 before truncation. A value of 20 was used for X_n because it provided fast training and high predictive ability.

The correlation coefficients $p(x_k, x_l)$ between each pair of descriptors was calculated for the $(69^2 - 69)/2 = 2346$ pairs. Of these, 24 (1 %) had p values greater than .90, of which 11 were correlations between adjacent BED and WED environment descriptors. A total of 756 (32 %) had p values in the range $-.1 < p < .1$. The average p value for all 2346 pairs was .07.

Results

Training consisted of choosing 20 random sets of 209 compounds from the total data set of 219 compounds using a random selection routine. This resulted in 20 sets of 209 known and 10 unknowns each. These sets were then used to train two binary pattern classifiers: one to make the classification tranquilizers versus nontranquilizers and the other to make the classification sedatives versus nonsedatives. Having two classification rules allows the inclusion in the data set of molecules that are neither sedatives nor tranquilizers or that have been reported as showing both activities. After an individual classifier was trained on 209 compounds, the 10 unknowns were predicted. The overall predictive ability was taken to be the average percentage of all 200 unknowns that were correctly classified after all 20 trainings.

Since the classifier develops a decision surface to separate the classes, the predictive ability of the classifier is dependent to a large extent on the number of members in the two classes of the training set that are nearest to the opposite class. The ability of the classifier to predict a compound having a rather small perturbation in properties from others in the data set is most accurately measured by sequentially removing one member from the data set, training, and predicting using the removed member as an unknown. This is known as the leave-one-out procedure (10,11). The rationalization for leaving 10 out was to provide a measure closely related to that of the leave-one-out procedure, while reducing the amount of computer time necessary to obtain such a measurement. The average percentage predictive ability is then an approximate measure of the success that a classifier developed from all 219 compounds would have in correctly classifying an unknown not contained

Application to Psychotropic Agents

Table 6.4 Results of Training and Predicting Using NDATA and NAVDATA

	NDATA			NAVDATA	
Threshold Value (Z)	Percent Prediction $Z = 0$	Percent Prediction $Z > 0$	Threshold Value (Z)	Percent Prediction $Z = 0$	Percent Prediction $Z > 0$
0	89.5	—	0	86.0	—
0.5	90.5	92.0	0.5	87.5	90.0
			1.0	89.0	90.0
			1.5	87.8	90.5
			2.0	88.3	90.0

in the original data set. This ability depends on the extent to which the data set is representative of that compound. As in any learning process, the smaller the perturbation from previous experience the more likely the chance for a successful classification.

The entire procedure was repeated a number of times using different threshold values. Table 6.4 summarizes the results of these studies for the tranquilizer versus nontranquilizer pattern classifier. It is seen that increasing the threshold causes an increase in predictive ability. The NDATA generally yielded better prediction than the NAVDATA; however, the NAVDATA normally required fewer feedbacks during training. In every case reported in Table 6.3, training was performed with the threshold value, Z, shown, and prediction was performed with a zero threshold and with the same threshold as used during training. The two predictive ability results are reported separately.

Generally it would not be thought that all the descriptors were of equal value in the training of the classifiers. To determine which of the 69 descriptors were of the greatest importance, both weight–sign and variance feature selection were employed.

For the weight–sign method, each of the 20 randomly selected training sets was used for training using a threshold of 1.75, and the remaining 10 drugs were predicted. The average of all 20 predictions was taken as a measure of the predictive ability prior to feature selection. Then the entire data set was used for training two weight vectors using one of the initializations given in Table 6.5 and with the sequence of the compounds randomly scrambled. Any descriptor whose weight vector component sign did not agree for the two weights was discarded. The process was then repeated until no further features could be eliminated. Then the predictive ability was tested using the

Table 6.5 Weight Vector Initializations

1. $w_j = 0$, $j = 1, 2, \ldots, (n-1)$; $w_n = 1$
2. $w_j = 0$, $j = 1, 2, \ldots, (n-1)$; $w_n = -1$
3. $w_j = 1$, $j = 1, 2, \ldots, (n-1)$; $w_n = 1$
4. $w_j = 1$, $j = 1, 2, \ldots, (n-1)$; $w_n = -1$
5. $w_1 = 1$; $w_j = 0, j = 2, 3, \ldots, n$
6. $w_1 = -1$; $w_j = 0, j = 2, 3, \ldots, n$
7. $w_j = -1$, $j = 1, 2, \ldots, (n-1)$; $w_n = 1$

same procedure used before feature selection. Results obtained for the seven independent implementations of the weight–sign feature selection procedure for the tranquilizer classification are shown in Table 6.6. The columns labeled No. Not Predicted give the total number of unknowns for all 20 trainings that were not classified because their dot products fell inside the dead zone, that is, $-Z < s < Z$. It is seen that in every case the predictive ability was higher after the unnecessary descriptors have been discarded. Also note that the total number of compounds not predicted was never greater for the feature selected data than the nonfeature selected data, and was in fact smaller for the former.

Since none of the original 69 descriptors alone has the ability to correctly classify all the data, some combination of features must be used to attain separation. The results of the weight–sign feature selection procedure can be

Table 6.6 Results of Weight–Sign Feature Selection Using the Seven Randomly Chosen Sets of Data

	69 Descriptors			After Feature Selection		
Percent Prediction[a]	No. not Predicted	Average Feedbacks	Percent Prediction[a]	No. not Predicted	Average Feedbacks	Descriptors Remaining
1. 85.08	5	386	88.94	2	340	40
2. 89.44	2	491	92.00	1	331	40
3. 86.00	1	266	90.00	0	274	38
4. 86.00	0	257	87.00	0	304	44
5. 85.89	1	300	91.44	1	187	35
6. 88.50	0	289	90.00	0	262	40
7. 86.00	0	280	87.50	0	223	34

[a] Training and prediction were done using $Z = 1.75$.

Table 6.7 Results of Feature Selection for Seven Randomly Chosen Sets of Data

Fraction of Times Retained	Descriptors Retained
7/7	9, 15, 22, 26, 37, 39, 42, 45, 48, 49, 61
6/7	1, 5, 6, 10, 24, 28, 31, 35, 52, 57, 58, 60
5/7	7, 11, 13, 20, 32, 47, 54, 59
4/7	2, 12, 14, 18, 21, 25, 34
3/7	23, 29, 31, 33, 44, 46, 51, 53, 56
2/7	3, 8, 16, 27, 30, 38, 50, 55, 65
1/7	4, 19, 34, 40, 41, 63, 66, 67
0/7	17, 62, 68, 69

used to group the descriptors as to relative importance in performing the classification. Table 6.7 gives the descriptors as a function of the fraction of the time they were retained during the seven feature selection operations. Eleven descriptors (16%) were retained every time, while only four descriptors (6%) were eliminated in all the runs. The remaining 54 descriptors (78%) were intermediate in importance.

The exact sequence in which descriptors are eliminated and, therefore, the identity of the remaining features depend on how the weight–sign feature selection procedure is implemented. For each of the seven weight vector initializations, a different set of features was eliminated, and different average predictive abilities were obtained, as shown in Table 6.8. In all cases the data remained linearly separable. This behavior is possible because there exists more than one unique set of descriptors that afford linear separability. Further weight–sign feature selection could be done by eliminating descriptors retained infrequently and implementing the entire procedure on the reduced data set.

Rather than reapplying the weight–sign procedure, the variance method of feature selection was applied. Thus a comparison could be made of the relative ability of this method to reduce the total number of features.

The procedure for application of the variance method is given in Chapter 4. The lower section of Table 6.8 shows the results obtained by applying variance feature selection to the data. The first iteration reduced the dimensionality from 69 to 31, and the second iteration further reduced the number to 23. A recognition rate of 100% was retained. A summary of the results from the two methods is given in Table 6.8. Clearly, the variance method offers a superior method of feature selection.

Table 6.8 Comparison of Results Obtained Using Several Methods of Feature Selection

Method Used	Initial Number of Descriptors	Final Number of Descriptors	Weight Vector Initialization	Iterations
Weight–sign analysis	69	38	1	8
	69	44	3	5
	69	34	4	9
	69	40	5	7
	69	40	6	9
Variance feature selection	69	31	Initial $x_n = 5$; increment $= 300$	
	31	23	Initial $x_n = 305$; increment $= 300$ *Final features retained*: 2, 7, 9, 10, 20, 22, 24, 26, 31, 32, 33, 37, 39, 42, 45, 46, 47, 48, 52, 55, 58, 61, 63	

Discussion

The data set used here represents a fairly diverse set of compounds, not all of which have their mode of action in the same area of the biosystem. It is rather unlikely that the exact mechanisms of these compounds' actions are all intrinsically related, although all are CNS agents. However, many other factors come into play in the relation of structure to function. Solubility and those parameters that affect solubility, geometric considerations, electronic properties, and so forth, also affect the observed activity. Given the sum total of this, similarities may exist that distinguish one class of compounds from the other.

The development of relevant descriptors and an efficient means of feature selection is then the crux of this problem. As shown in Table 6.7, any single feature selection was able to reduce the features required for separation by at least 40% while maintaining or increasing the predictive ability and reducing the number of compounds that fell into the deadzone. It is further shown (Table 6.8) that only 23 features ultimately were necessary to define these relationships.

Studies show that descriptors such as the number of nitrogens, total bond length, and various fragment and topological descriptors ranked high in their importance for discrimination as applied to this data. While this does not indicate that these parameters relate to the properties of sedation and tranquilization by a cause and effect relationship, it does imply that certain similarities exist within each class of compounds such that an approximation of biological effect can be made on the basis of structure.

APPLICATION TO BARBITURATES

The present study is concerned with the utility of pattern recognition techniques for differentiating between barbituric acids according to the duration of their effect. As in the previous study, classification is based entirely on information available from a standard two-dimensional structural representation of the molecule. The purpose of this study is to demonstrate the utility of pattern recognition techniques in developing rules to classify the compounds according to duration of the compounds' effect. We also demonstrate the ability of these techniques to isolate the features responsible for defining the discriminant function. An estimation of the reliability of the rules that are developed is included.

Data Set

The set of compounds used in the present study consists of 160 5,5'-substituted barbiturates selected from a standard reference (12). These compounds range in molecular weight from 172 to 276 and have duration times ranging from 10 to 1600 minutes. The method of administration was either intraperitoneal or subcutaneous, using mice, rats, or rabbits as test animals. The compounds are given in Table 6.9. The fact that this data set is heterogeneous in mode of administration and test species, but that it can still be dealt with using pattern recognition methods, illustrates one of the strengths of the approach. The methods employed in pattern recognition often make it possible to study incomplete, ill-defined, or otherwise imperfect sets of compounds, while many other more rigorous methods demand better quality data sets. The success enjoyed in analyzing this barbiturate data set in the present study is meant to indicate how pattern recognition methods can be used, and it is not meant to point out how to synthesize new barbiturates.

The compounds were grouped into classes according to the duration of depressant effect. These classes were formed by dividing the duration time expressed in minutes by 10. The resulting class designation was rounded up if the remainder was 5 or greater and it was rounded down otherwise. Thus a compound whose duration time was 227 minutes would be placed in class 23, whereas a compound having a duration time of 223 minutes would be placed

Table 6.9 Compounds Forming the Data Set

R	R'	Duration (minutes)
CH_3-	1. $(CH_3)_3CCH-$	580
	2. $CH_3(CH_2)_5-$	260
	3. $CH_3(CH_2)_3CH(CH_3)-$	227
	4. $CH_3(CH_2)_3CH(CH_3CH_2)CH_2-$	223
	5. $H_2C=C(CH_3)-$	60
	6. $CH_3CH=C(CH_3)-$	120
	7. $CH_3CH_2HC=C(CH_3)-$	60
	8. $CH_3HC=C(CH_3CH_2)-$	60
	9. $CH_3(CH_2)_2HC=C(CH_3)-$	60
	10. $(CH_3)_2CHC=C(CH_3)-$	36
	11. $CH_3(CH_2)_3HC=C(CH_3)-$	24
	12. $CH_3CH_2SCH_2-$	330
	13. $CH_3(CH_2)_3SCH_2-$	150
CH_3CH_2-	14. CH_3CH_2-	1400
	15. $CH_3CH_2CH_2-$	1140
	16. $CH_3CH(CH_3)-$	1520
	17. $CH_3CH_2CH_2CH_2-$	450
	18. $CH_3CH(CH_3)CH_2-$	540
	19. $CH_3CH_2CH(CH_3)-$	600
	20. $CH_3CH_2CH_2CH_2CH_2-$	220
	21. $CH_3CH_2CH(CH_3)CH_2-$	190
	22. $(CH_3)_3CCH_2-$	200
	23. $CH_3CH_2CH(CH_3CH_2)-$	300
	24. $CH_3(CH_2)_5-$	45
	25. $CH_3(CH_2)_2CH(CH_3)CH_2-$	210
	26. $CH_3CH_2C(CH_3)_2CH_2-$	60
	27. $CH_3(CH_2)_3CH(CH_3)-$	90
	28. $CH_3CH_2CH(CH_3CH_2)CH_2-$	300
	29. $CH_3(CH_2)_6-$	120
	30. $(CH_3)_2CHCH_2CH(CH_3)CH_2-$	54
	31. $(CH_3)_2CH(CH_2)_2CH(CH_3)-$	50
	32. $CH_3CH_2CH(CH_3)CH_2CH(CH_3)-$	74
	33. $CH_3(CH_2)_2CH(CH_3CH_2CH_2)-$	81
	34. $CH_3(CH_2)_2CH(CH_3)CH_2CH_2CH_2-$	60
	35. $CH_3CH_2CH(CH_3)CH_2CH(CH_3)CH_2-$	60
	36. $CH_3(CH_2)_3CH(CH_3CH_2)CH_2-$	75
	37. $CH_3(CH_2)_4CH(CH_3CH_2)-$	60
	38. $CH_3(CH_2)_5CH(CH_3)-$	150
	39. $CH_3(CH_2)_2CH(CH_3)CH(CH_3CH_2)CH_2-$	240
	40. $(CH_3)_2CH(CH_2)_2CH(CH_3CH_2)CH_2-$	120
	41. $H_2C=CH-$	288

Table 6.9 (*Continued*)

R	R′	Duration (minutes)
	42. H$_2$C=C(CH$_3$)—	150
	43. CH$_3$CH$_2$HC=CH—	18
	44. CH$_3$HC=C(CH$_3$)—	180
	45. (CH$_3$)$_2$C=CH—	240
	46. CH$_3$(CH$_2$)$_2$HC=CH—	96
	47. CH$_3$CH$_2$HC=C(CH$_3$)—	24
	48. (CH$_3$)$_2$CHHC=CH—	12
	49. CH$_3$HC=C(CH$_3$CH$_2$)—	42
	50. CH$_3$(CH$_2$)$_2$HC=C(CH$_3$)—	72
	51. CH$_3$(CH$_2$)$_3$HC=C(CH$_3$)—	6
	52. CH$_3$CH$_2$HC=C(CH$_3$CH$_2$CH$_2$)—	6
	53. H$_2$C=CHCH(CH$_3$)—	720
	54. H$_2$C=C(CH$_3$)CH$_2$—	326
	55. CH$_3$HC=CHCH$_2$	372
	56. CH$_3$CH$_2$OCH(CH$_3$)—	460
	57. CH$_3$(CH$_2$)$_2$OCH(CH$_3$)—	150
	58. CH$_3$(CH$_2$)$_3$OCH(CH$_3$)—	150
	59. (CH$_3$)$_3$CCH$_2$OCH(CH$_3$)—	75
	60. CH$_3$CH$_2$OC(H$_2$C)—	200
	61. (CH$_3$)$_3$CCH$_2$OC(CH$_2$)—	63
	62. CH$_3$(CH$_2$)$_2$SCH$_2$—	59
	63. (CH$_3$)$_2$CHSCH$_2$—	139
	64. H$_2$C=CHCH$_2$SCH$_2$	117
	65. CH$_3$(CH$_2$)$_3$SCH$_2$—	66
	66. CH$_3$(CH$_2$)$_4$SCH$_2$—	75
	67. (CH$_3$)$_3$CCH$_2$SCH$_2$—	37
	68. CH$_3$(CH$_2$)$_2$CH(CH$_3$)SCH$_2$—	62
	69. CH$_3$(CH$_2$)$_5$SCH$_2$—	15
	70. (CH$_3$CH$_2$)$_2$CHCH$_2$SCH$_2$—	22
	71. CH$_3$CH$_2$SCH(CH$_3$CHCH$_3$)—	12
	72. CH$_3$(CH$_2$)$_3$SCH(CH$_3$)—	34
	73. CH$_3$(CH$_2$)$_3$SCH(CH$_3$CH$_2$)—	52
	74. CH$_3$(CH$_2$)$_4$SCH(CH$_3$)—	28
	75. (CH$_3$)$_3$CCH$_2$SCH(CH$_3$)—	41
	76. CH$_3$(CH$_2$)$_2$CH(CH$_3$)—	180
CH$_3$CH$_2$CH$_2$—	77. CH$_3$CH$_2$CH$_2$CH$_2$CH$_2$—	4
	78. CH$_3$CH$_2$CH(CH$_3$)CH$_2$—	165
	79. CH$_3$(CH$_2$)$_5$—	1
	80. CH$_3$(CH$_2$)$_6$—	15

(*Continued*)

Table 6.9 (*Continued*)

R	R'	Duration (minutes)
	81. $CH_3HC{=}CH{-}$	60
	82. $CH_3CH_2HC{=}CH{-}$	18
	83. $(CH_3)_2CHHC{=}CH{-}$	18
	84. $H_2C{=}C(CH_3){-}$	168
	85. $CH_3HC{=}C(CH_3){-}$	30
	86. $CH_3CH_2HC{=}C(CH_3){-}$	18
	87. $CH_3HC{=}C(CH_3CH_2){-}$	24
	88. $H_2C{=}CHCH(CH_3){-}$	420
	89. $H_2C{=}C(CH_3)CH_2{-}$	300
	90. $CH_3HC{=}CHCH_2{-}$	120
	91. $CH_3CH_2OCH(CH_3){-}$	162
	92. $CH_3CH_2SCH_2{-}$	150
	93. $CH_3(CH_2)_3SCH_2{-}$	76
	94. $CH_3(CH_2)_3SCH(CH_3){-}$	35
	95. $(CH_3)_2CHCH_2SCH(CH_3){-}$	45
$(CH_3)_2CH{-}$	96. $(CH_3)_2CHCH_2{-}$	25
	97. $CH_3HC{=}CH{-}$	36
	98. $CH_3CH_2HC{=}CH{-}$	36
	99. $CH_3(CH_2)_2HC{=}CH{-}$	18
	100. $(CH_3)_2CHHC{=}CH{-}$	12
	101. $CH_3CH_2HC{=}C(CH_3){-}$	18
	102. $CH_3C{=}C(CH_3CH_2){-}$	18
	103. $H_2C{=}CHCH(CH_3){-}$	210
	104. $CH_3HC{=}CHCH_2{-}$	200
	105. $CH_3CH_2SCH_2{-}$	86
	106. $CH_3(CH_2)_3SCH_2{-}$	38
$CH_3(CH_2)_3{-}$	107. $CH_3CH_2CH(CH_3){-}$	16
	108. $(CH_3)_3C{-}$	1
	109. $CH_3HC{=}CH{-}$	12
	110. $CH_3CH_2HC{=}CH$	18
	111. $H_2C{=}C(CH_3){-}$	90
	112. $CH_2HC{=}C(CH_3){-}$	60
	113. $H_2C{=}CHCH(CH_3){-}$	110
	114. $CH_3HC{=}CHCH_2{-}$	40
	115. $(CH_3)_2C{=}CHCH_2{-}$	30
	116. $CH_3CH_2OCH(CH_3){-}$	120
	117. $CH_3CH_2SCH_2{-}$	74
	118. $CH_3(CH_2)_3SCH_2{-}$	95

Table 6.9 (*Continued*)

R	R'	Duration (minutes)
$H_2C=CH-$	119. $CH_3CH_2CH_2CH_2-$	288
	120. $(CH_3)_3CCH-$	192
$H_2C=CH(CH_3)-$	121. $H_2C=CHCH_2$	102
	122. $(CH_3)_2CHCH_2-$	90
	123. $CH_3CH_2CH_2CH_2CH_2-$	30
	124. $(CH_3)_3CCH-$	18
$CH_3HC=C(CH_3)-$	125. $H_2C=CHCH_2-$	30
$H_2C=CHCH_2-$	126. $CH_3(CH_2)_3CH(CH_3)-$	108
	127. $H_2C-CHCH(CH_3)-$	456
	128. $CH_3CH_2OCH(CH_3)-$	300
	129. $CH_3CH_2OC(CH_2)-$	300
	130. $CH_3(CH_2)_2OCH(CH_3)-$	204
	131. $(CH_3)_3CCH_2OCH_2-$	900
	132. $H_2C=C(CH_3)CH_2-$	380
	133. $CH_3CH_2SCH_2-$	164
	134. $CH_3(CH_2)_2SCH_2-$	117
	135. $CH_3(CH_2)_3SCH_2-$	123
	136. $CH_3(CH_2)_3SCH(CH_3)-$	34
	137. $(CH_3)_2CHCH_2-$	162
	138. $(CH_3)_3CCH_2-$	96
	139. $H_2C=CHCH_2-$	880
	140. $(CH_3)_2CH-$	720
	141. $CH_3(CH_2)_2CH(CH_3)-$	150
$CH_3HC=CHCH_2-$	142. $(CH_3)_3CCH-$	40
	143. $CH_3(CH_2)_2CH(CH_3)-$	66
	144. $CH_3CH_2CH(CH_3)-$	120
	145. $(CH_3)_3CHCH_2-$	45
$(CH_3)_2C=CHCH_2-$	146. $(CH_3)_2C=CHCH_2$	70
	147. $CH_3CH_2CH(CH_3)-$	120
$CH_3CH_2OCH(CH_3)-$	148. $(CH_3)_3CCH_2-$	102
	149. $CH_3(CH_2)_2CH(CH_3)-$	108
$CH_3(CH_2)_2OCH(CH_3)-$	150. $CH_3CH_2CH_2CH(CH_3)-$	300

(*Continued*)

Table 6.9 (*Continued*)

R	R'	Duration (minutes)
CH_3SCH_2-	151. $(CH_3)_2CHCH_2-$	108
$CH_3CH_2SCH_2-$	152. CH_3CH_2-	143
	153. $(CH_3)_2CHCH_2-$	81
	154. $CH_3CH_2CH(CH_3)-$	61
	155. $(CH_3)_3CCH_2-$	8
	156. $CH_3(CH_2)_2CH(CH_3)-$	35
$CH_3CH_2SCH(CH_3)-$	157. $CH_3(CH_2)_5-$	12
$H_2C{=}CHCH_2SCH(CH_3)-$	158. $(CH_3)_2CHCH_2-$	28
$CH_3(CH_2)_3SCH_2-$	159. $(CH_3)_2CHCH_2-$	78
	160. $CH_3CH_2CH(CH_3)-$	69

in class 22. Compounds with a duration greater than 650 minutes were placed in class 65. This resulted in a total of 65 different classes, which are distributed as shown in Figure 6.1.

Four types of descriptors were employed for these studies: numeric fragment descriptors, substructural descriptors, environmental descriptors, and molecular connectivity descriptors. The descriptors were generated using the automated descriptor packages described in Chapter 3. The initial set of descriptors used is given in Table 6.10.

While the nature of the atom, bond, and substructural descriptors is obvious, further comment is necessary regarding the environment and molecular connectivity descriptors. The environment descriptor takes into account how different parts of the molecule are connected by providing a measure of the local environment of a single atom fragment. This is accomplished by combining the fragment's first and second nearest neighbors and their bonds into a single parameter that reflects the chemical environment around the fragment. This study employs three types of environment descriptors: the bond environment descriptor (BED), which uses only the number of bonds to calculate a descriptor value; the weighted environment descriptor (WED), which uses the type of bond in the calculation; and the augmented environmental descriptor (AED), which uses both the type of atom and type of bond in the calculation.

Application to Barbiturates

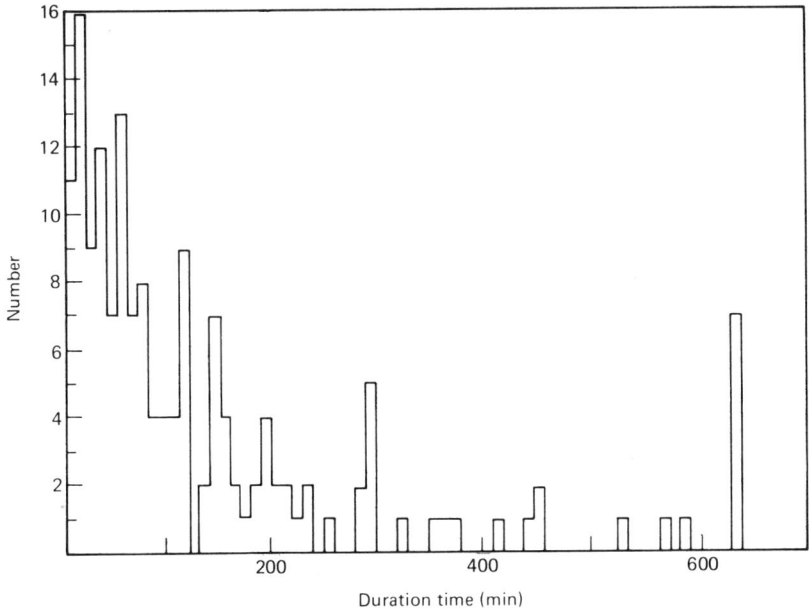

Figure 6.1 Histogram of barbiturate duration times.

The molecular connectivity descriptor provides a measure of the connectivity for the entire molecule. The concept was developed by Randić (13) and was used in structure-activity studies by Kier et al. (14–17). Both have shown a number of correlations between the molecular connectivity and several different physical parameters (18). The connectivity index is calculated directly from the connection table representation of the molecules as described in Chapter 3. We have used as descriptors the simple index, the index corrected for rings, and the square of these indices. Ring correction was accomplished by subtracting from the simple index a value equal to the average of the contributions from all bonds that are contained in a ring. The descriptors were then multiplied by 10 and truncated to integer values.

Thus the data set consists of 160 compounds each coded with 47 descriptors. In no case does any one descriptor, or any binary combination of descriptors, contain sufficient information to successfully classify the data. Preprocessing of the raw data prior to classification consisted of autoscaling so that each descriptor had an average of zero and a standard deviation of 127. This allowed the data to be truncated to integer values with a negligible loss of precision (recalculation after truncation yields a standard deviation of 127 and a mean of 0 ± 0.17).

Table 6.10 Molecular Structure Descriptors

Atom and Bond Descriptors

1. Number of atoms
2. Number of bonds
3. Number of carbon atoms
4. Number of nitrogen atoms
5. Number of oxygen atoms
6. Number of single bonds
7. Number of double bonds
8. Length[a]

Environment Descriptors

Atom Centered Fragment	General[b]	Cyclic
9–11. CH_3-	1, 2, 3	
12–14. $-CH_2-$	1, 2, 3	
15–17. $-CH-$	1, 2, 3	
18–23. $-C-$	1, 2, 3	1, 2, 3
24–26. $O=$	1, 2, 3	
27–29. $-HC=$	1, 2, 3	
30–35. $C=$	1, 2, 3	1, 2, 3

Substructural Descriptors

36. CH_3CH_2- 37. $-CH(CH_3)CH_2-$ 38. CH_3-
39. $-CH_2-$ 40. $-CH_2CH_2-$ 41. $CH_3CH_2CH_2-$
42. $-CH-$ 43. $-HC=$

Molecular Connectivity Descriptors[c]

44. MC1 45. MC2 46. MC3
47. MC4

[a] Length = 4* (number of single bonds) + 2* (number of double bonds).
[b] 1 = BED, 2 = WED, 3 = AED.
[c] MC1 = simple index, MC2 = ring corrected index, MC3 = the square of the simple index, MC4 = the square of the ring corrected index.

The learning machine requires that a constant valued descriptor be added to the data set. In the present studies a value of 250 was used because it provided for fast training and high predictive abilities. This parameter is discussed further in Chapter 4.

Results

The duration of the barbiturate depressant effect is highly dependent on the conditions under which a compound is tested. The data compiled for this study represent a series of studies on different animals, at different laboratories, and it is expected that there will be a large degree of variation within the data. However, several series of compounds are present that were tested as a group and, therefore, trends in the duration that correlate to structural alterations may exist.

To account for these variations, any one classifier will develop a discriminant that answers the question, Is the duration time less than x minutes?, where compounds within 30 minutes of this duration time are not used to develop the discriminant. Using the class designations formed as noted previously, there are 61 possible ways of forming two clusters (longer or shorter than duration x) such that a gap of three classes lies between the clusters. Our initial studies showed that after the removal of compounds 5, 29, 38, 44, 121, and 150, it was possible to develop discriminants for each of these 61 cluster sets; however, only three such sets are used here to demonstrate the method. The six compounds that were removed are discussed later.

Set I assigned all members in classes 1 through 10 to the short duration cluster and classes 14 through 65 to the long duration cluster, set II assigned classes 1 through 20 to the short duration cluster and 24 through 65 to the long duration cluster, and set III assigned classes 1 through 24 to the short duration cluster and 28 through 65 to the long duration cluster. Using these three sets discriminants can be developed to classify compounds as having a duration of less than 100 minutes, less than 200 minutes, or less than 240 minutes. Compounds belonging to a class of longer duration would not be assigned to any of these duration regions.

One method of assessing the reliability of these discriminant functions is to subdivide each of the three sets so that each successive group contains more members in the prediction set and fewer members in the training set. These groups can be used to estimate the predictive ability and to determine which descriptors support the discriminant function's ability to separate the cluster of short duration barbiturates from those of long duration. If within each set of clusters the descriptors selected vary significantly and the predictive abilities are quite different, it would be clear that no clusters actually existed and that no relation between the structure and duration was found.

Results for the predictive ability tests and for feature selection using descriptors 1 through 43 are shown in Table 6.11. The portion of the data placed in the prediction set is indicated at the top of each column. Equal percentages of the long and short duration clusters were taken to form these sets. The remaining members were placed in the training set. Ten such sets were formed for each percentage group. The highest predicting of these sets was used to select the features responsible for the discriminant's ability to classify the data.

Feature selection was accomplished using the variance feature selection method described in Chapter 4. Those descriptors that were retained are indicated by the symbol ×. The predictive ability is the average for all 10 sets before and after the feature selection process. Total refers to the results of feature selection using all the members of each cluster. Reference refers to the results for the one prediction and training set used in the feature selection

Table 6.11 Comparison of the Descriptors Retained for each Cluster Set

Descriptor No.	Set I				Set II				Set III			
	Total	10%	15%	20%	Total	10%	15%	20%	Total	10%	15%	20%
1									×			
2												×
3	×											
5		×	×	×	×	×		×	×	×	×	×
6	×	×	×				×	×	×		×	
7	×	×	×	×	×	×	×	×		×	×	×
8						×				×		
9			×	×								
10		×	×				×					
11					×	×			×			
12	×											
14			×									
15	×	×	×	×	×	×	×	×	×			
16					×	×	×	×	×		×	
17								×				
19										×		
20					×							
21						×						
23					×		×		×			
27	×	×	×	×							×	

Application to Barbiturates

Table 6.11 (*Continued*)

Descriptor No.	Set I				Set II				Set III			
	Total	10%	15%	20%	Total	10%	15%	20%	Total	10%	15%	20%
28		×	×							×		×
30								×	×	×		
32	×	×		×				×	×	×	×	×
33		×			×	×	×					
34			×	×	×		×		×	×	×	×
35	×			×	×	×	×		×	×	×	×
36	×	×	×	×								
37								×		×	×	×
38				×					×	×	×	×
39	×	×		×	×	×						
40					×			×	×	×	×	×
41		×										
42					×	×	×	×				
43	×	×			×	×	×	×				
Reference[a]												
Initial	—	100	95.5	96.6	—	100	100	96.8	—	100	95.4	96.9
Final	—	100	100	100	—	100	100	93.6	—	100	95.4	96.9
Total set[b]												
Initial	—	92.0	88.2	89.0	—	91.9	90.4	91.0	—	91.9	93.3	93.4
Final	—	92.7	91.8	92.4	—	94.4	93.0	92.9	—	95.0	95.4	95.3

[a] Predictive ability for feature selection reference set.
[b] Average predictive ability for the 10 prediction sets within each percentage group.

process. The members of this prediction set were never used to develop the discriminant function and therefore represent total unknowns.

Note that the molecular connectivity descriptors (numbers 44 to 47) were not included in these initial studies. These were omitted to keep the ratio of compounds to descriptors above 3:1. This is necessary to ensure that a nontrivial discriminant function is developed (19). To include descriptors 44 through 47, a reduced set of the first 43 descriptors was chosen for each set by pooling those descriptors from Table 6.11 that were selected three or more times. Using these as the initial descriptors, each set of clusters was feature selected by the variance method. The resulting descriptors represent a minimum set, that is, if any of the selected descriptors are excluded from the training process, a linear discriminant function that separates the data cannot be developed. Descriptors 44 through 47 were then added to these reduced sets and each was once again subject to variance feature selection. The descriptors that were ultimately selected for each set of clusters are shown in Table 6.12.

The best measure of predictive ability is obtained by leaving out one compound and using the remaining compounds as the training set. The surface developed from the training set is used to predict the cluster into which the remaining (and therefore unknown) compound belongs. This procedure is continued until each member of the data set has been left out of the training set once. The predictive ability is the number of correct classifications divided by the total number of classifications. For a finite set, this method is considered the most unbiased estimator of predictive ability. Approximations to this measurement can be made by repeating the process several times using a larger prediction set. The predictive abilities in Table 6.12 were estimated using this leave-one-out procedure.

Table 6.13 gives the mean value, autoscale factors, and the weight vector for the set I discriminant. To predict whether an unknown has an activity of less than 100 minutes, the descriptors from Table 6.10 are calculated and these values are scaled by subtracting the mean value for that descriptor, multiplying the result by the normalizing factor, and truncating the results to integers. The result is a nine component vector. Using a value of 250 as the tenth component, the dot product of this vector and the weight vector is calculated. If the sign of the dot product is positive, the activity is less than 100 minutes.

The calculation for the barbiturate having R = ethyl, $R' = sec$-pentyl is given as an example. This compound is not part of the original data set and therefore constitutes an unknown. The duration is reported to be 180 minutes (20). Calculation of the descriptors in set I yields the vector $X = (3, 3, 19, 17, 0, 106, 47, 2, 71)$. Normalizing this vector yields $X_n = (-41, -116, 9, 117, -91, -76, 105, 105, -42)$. Adding the extra component and calculating the dot products yields -30.6. Since the sign of the dot product is negative, the duration is estimated as being greater than 100 minutes. It should be noted that discriminant functions can be used to predict activities and unknowns, as has been done here, using a simple calculation that can be done with a desk calculator (if the descriptors can be hand calculated).

Discussion

The fact that discriminants could be developed successfully for a data set as diverse and heterogeneous as this one indicates that information concerning duration of depressant effect is contained in the structure of these compounds. While it is possible to develop discriminant functions that evidence chance correlations, the experiments performed indicate that such correlations are not responsible for the behavior of the discriminants.

Table 6.12 Descriptors Selected for Cluster Sets I, II, and III

Set I Atom and Bond Descriptors		Set II Atom and Bond Descriptors		Set III Atom and Bond Descriptors	
Substructural Descriptors	Environment Descriptors[a]	Substructural Descriptors	Environment Descriptors[a]	Substructural Descriptors	Environment Descriptors[a]
Number of oxygen atoms Number of double bonds		Number of oxygen atoms Number of double bonds		Number of oxygen atoms	
CH_3CH_2-	CH_3- (G, 2)	CH_3CH_2-	CH_3- (G, 3)	CH_3-	$-\overset{\mid}{H}C-$ (G, 1)
	$-\overset{\mid}{H}C-$ (G, 1)	$-\overset{\mid}{H}C-$	$-\overset{\mid}{H}C-$ (G, 2)	$-CH_2CH_2-$	$-HC=$ (G, 1)
	$\diagup C= \diagdown$ (G, 3) (C, 1)	$-HC=$	$\diagup C= \diagdown$ (C, 1) (C, 3)	$-CH(CH_3)CH_2-$	$\diagup C= \diagdown$ (G, 3) (C, 3)
	$-HC=$ (G, 1)				
Molecular Connectivity MC2		Molecular Connectivity MC4		Molecular Connectivity MC4	
Average predictive ability 93.8%		92.9%		93.7%	

[a] G, indicates that the descriptor was generated for every occurance of the indicated fragment. C indicates that the descriptor was generated for only those fragments which appeared in a ring. 1 indicates that the descriptor was a BED type. 2 indicates that the descriptor was a WED type. 3 indicates that the descriptor was an AED type.

Table 6.13 Weight Vector and Normalizing Factors for Cluster Set I

Descriptor Number	Mean Value	Mult/Sigma	Weight Vector
5	3.907	431.4560	−0.2197
7	3.527	220.8830	−0.4915
10	18.376	15.6835	0.0441
15	8.457	13.7709	−0.2994
27	5.527	16.4919	0.4415
32	112.648	11.4809	0.2787
33	45.994	104.7780	0.2545
36	1.333	158.7940	−0.1682
45	63.752	15.3582	0.5009
$N + 1$	250		0.0453

The reliability of relations found using nonparametric discriminant analysis is a function of the discrimination ability of the classifier. Discrimination ability is a measure of the classifier's ability to find a separating discriminant function. In each of the sets studied, the learning machine could find a discriminant that would separate short duration members from long duration members. If chance correlations were responsible for this separation ability, the features selected for the members in the training set would not support a separating discriminant for members of the training and prediction set. For each different training set in Table 6.11, the descriptors chosen were similar. Additional experiments showed these descriptors would support a discriminant that separated all the members of the training and prediction sets. Thus the same structural features that are intrinsic to the development of a discriminant for the training set are also intrinsic to the relations for the prediction set.

The predictive ability of a discriminant depends on how the discriminant was developed. The linear learning machine does not necessarily provide a discriminant that yields the best predictive ability. While a training set may be linearly separable, there are an infinite number of separating discriminants. Thus even though it is possible to use the descriptors selected by the training set to develop a discriminant for the prediction set, such a function does not necessarily perform well. The predictive ability is a gauge of how well a discriminant will classify the data not used in developing that function. When developed from several different training sets, a discriminant that develops

Application to Barbiturates

chance correlations would evidence low or variable predictive ability. The studies summarized in Table 6.11 show that decreasing the number of members used to develop the discriminant function does not substantially degrade its predictive performance.

The discriminants developed by the learning machine are seen to be quite general. Not only can they distinguish among different structures, but differences in duration between congeners are also described. Structures 17, 20 and 24 constitute a congeneric series of increasing alkyl chain length. The fact that discriminants could be developed for all of the 61 possible divisions of the data set indicates that such a series of compounds can be distinguished. Similarly, these descriptors can account for a branched series such as 16, 19, and 27. The duration of structural isomers such as compounds 24, 25, and 27 and between 53 and 54 is also described.

In this light it is interesting to investigate some of the six compounds that could not be accounted for using these descriptors. Compound 29 is seen to be a member of the series 14, 15, 17, 20, and 24 and its duration time might be expected to be less than that for compound 24. Its duration, however, deviates from the order implied by these compounds since it is unexpectedly large. Most likely, the differences in its activity can be attributed to changes in lipophilic properties due to the sizable side chain. Similarly, compound 38 belongs to the series 16, 19, and 27. Its duration also deviates from that implied by the other members in the series. Arguments similar to those above can be made for compounds 5 and 44, which do not fit the pattern followed by the remainder of the data set. The data set does not fully represent structures 121 and 150 and thus it is not surprising that they are always incorrectly classified.

The ability to quickly identify those compounds differing from the larger body of data is a useful property of this approach to structure–activity studies. Once these differences are identified, they can be used to gain further information concerning the action of these compounds.

The structural parameters used in these studies appear consistent with the observed properties of the barbiturates. The lack of any dominant structural feature indicates a lack of specificity for the receptor site with which the compounds interact. The descriptors chosen through feature selection indicate that properties of chain length and the extent of branching are the major influences on barbiturate duration. It has been suggested that the amount of shielding of the 5 position may be responsible for many of the lipophilic properties (21). Descriptors 32, 33, and 35 were included in the final sets of features for the three thresholds. These environment descriptors extend to the secondary position of R and R' and could conceivably account for this shielding. Similarly, the molecular connectivity descriptors provide information on the degree of branching, which in turn can be related to lipophilic properties.

While it was not our intention to develop a discriminant that can be used for all barbiturates, evidence of the utility of the discriminants that were developed was provided by prediction of the activity of the compound having R = ethyl, and R' = *sec*-pentyl. The duration time of this compound was correctly predicted to lie between 100 and 200 minutes by using the discriminant given in Table 6.13 in combination with those from sets III and IV. Clearly, once a discriminant has been developed and the descriptors generated, the actual prediction process is quite straightforward.

A question naturally arises concerning the possibility of using the parameters that arise from pattern recognition analysis to produce structures of a specific activity. A direct path to this goal is not possible as can be seen from Table 6.14. The table gives pertinent statistics for each feature used to define the duration of the barbiturates with respect to set I. Listed are the standard deviation of the descriptors and the numerical average of the descriptor values. The highest and lowest values give information concerning the range of descriptor values.

Note that although the average values for the two classes differ from each other, the standard deviation is larger than this difference. Therefore, the individual descriptors, while providing structural information, are not the sole indicators of activity. In the case of atom, bond, and substructural descriptors, the average value can be related directly to the structural composition of the molecule. However, average values for the environment and mole-

Table 6.14 Statistics for Final Set of Descriptors Selected for Cluster Set I

	Mean		Standard Deviation			
Descriptor No.	Compounds below Threshold[a]	Compounds above Threshold[a]	Compounds below Threshold[a]	Compounds above Threshold[a]	Highest Value	Lowest Value
2	3.52	3.54	0.56	0.61	5	3
5	3.04	3.18	0.24	0.38	4	3
10	19.91	16.19	9.00	7.55	41	0
12	8.56	8.34	9.26	9.30	36	0
27	6.50	4.15	9.20	5.40	29	0
32	114.96	109.35	13.20	8.78	141	102
33	46.16	45.77	1.17	1.33	49	43
36	1.37	1.28	0.82	0.78	3	0
MC2	66.24	60.21	9.70	8.00	81	42

[a] Number of compounds above threshold is 56, number below is 90.

Application to Barbiturates

cular connectivity descriptors are difficult or impossible to interpret, as their relation to the structure is complex.

It is clear that average values indicate only the relative presence of a particular descriptor and cannot be construed as indicating the amount necessary for activity. A case in point is descriptor 5, the number of oxygen atoms in the molecule. This number ranges between 3 and 4. The barbiturate ring accounts for 3 of these atoms. A value greater than 3 includes the number in the side chains. The fact that on the average, the class of longer acting molecules contains slightly more oxygen atoms does not imply that adding oxygen guarantees an increase in the activity. The number, placement, and chemical environment of an oxygen govern its effectiveness, not its mere presence. If the activity of a molecule is to be described by use of structural parameters, each must be viewed as a single contribution to, rather than the single indication of, that action.

Since structurally derived descriptors reflect the composition of the structure, they are interdependent. Changes in composition generally affect the value of several structural parameters simultaneously. Most notably affected are the environment and molecular connectivity descriptors. However, substructural content is also influenced by slight alterations in the structure. Such alterations affect the placement of the molecule in the space formed by its descriptors and therefore affect the results of classification. This dependence arises because the biological activity expressed as a function of the molecular structure is a vector representation of the descriptors for that structure. The discriminant developed from pattern recognition analysis can be thought of as a transform, which maps a structure vector onto one of the two cluster regions. The reverse mapping cannot be accomplished directly. Alternately, structural descriptors can be viewed as indicating the electronic, steric, and lipophilic properties of a molecule. No one descriptor is an effective gauge of all these properties. Each is a component in their description. Knowing these properties does not allow the direct construction of active molecules.

Although the parameters used in the pattern recognition analysis cannot be used to directly construct active molecules, they can be used to predict the effectiveness of hypothetical structures. This offers a pragmatic aid in the synthesis problem, that is, given a choice of molecules that appear to be equally likely candidates for synthesis, how does one optimize the chances of synthesizing the most active. If a data base exists that details past successes and failures, then a plausible solution is to use the data base to develop rules that estimate the activity of a candidate structure. Since pattern recognition develops rules that define "similarity," then application of the methods as described in the preceding sections will aid in the synthesis decision.

These techniques could also prove useful in large scale prescreening. The derivation of structural parameters is rapid enough to allow several thousand prospective structures to be described and tested using a discriminant developed from a set of compounds known to be active. Those compounds that the discriminant notes as being the highest acting can then be considered for further testing. Prediction results from the sets studied indicate that such classifiers can perform with a high degree of reliability.

The ultimate purpose in the study of effects of structural alterations on biological action is to produce new, more effective compounds. Useful tools are those that produce information pertinent to this goal. Historically, the chemist has used a structural diagram of the molecule as a gauge for altering the structure. The number and diversity of active compounds attest to the success of this approach. As the effectiveness of this approach is dependent on the judgment of the chemist, use of mathematical techniques to augment these judgments may well increase the effectiveness of this procedure.

REFERENCES

1. C. Hansch, S. Unger, and A. B. Forsythe, Strategy in Drug Design. Cluster Analysis as an Aid in the Selection of Substituents, *J. Med. Chem.*, **16**, 1217 (1973).
2. K. L. H. Ting, R. C. T. Lee, G. W. A. Milne, M. Shapiro, and A. M. Guarino, Applications of Artificial Intelligence: Relationships between Mass Spectra and Pharmacological Activity of Drugs, *Science*, **180**, 417 (1973).
3. B. R. Kowalski and C. F. Bender, The Application of Pattern Recognition to Screening Prospective Anti-Cancer Drugs. Adenocarcinoma 755 Biological Activity Test, *J. Am. Chem. Soc.*, **96**, 916 (1974).
4. K. C. Chu, R. J. Feldmann, M. B. Shapiro, G. F. Hazard, Jr., and R. I. Geran, Pattern Recognition and Structure–Activity Relationship Studies. Computer-Assisted Prediction of Anti-Tumor Activity in Structurally Diverse Drugs in an Experimental Mouse Brain Tumor System, *J. Med. Chem.*, **18**, 539 (1975).
5. A. Cammarata and G. K. Menon, Pattern Recognition. Classification of Therapeutic Agents According to Pharmacophores, *J. Med. Chem.*, **19**, 739 (1976).
6. G. K. Menon and A. Cammarata, Pattern Recognition II: Investigation of Structure–Activity Relationships, *J. Pharm. Sci.*, **66**, 304 (1977).
7. F. Darvas, Application of the Sequential Simplex Method in Designing Drug Analogs, *J. Med. Chem.*, **17**, 799 (1974).
8. S. A. Hiller, U. C. Golender, A. B. Rosenblit, L. A. Rastrigin, and A. B. Glaz, Cybernetic Methods of Drug Design I. Statement of the Problem—The Perception Approach, *Comp. Biomed. Res.*, **6**, 411 (1973).
9. E. Usdin and D. H. Effron, *Psychotropic Drugs and Related Compounds* 2nd ed., DHEW Pub. No. (HSM) 72-9074, 1972.
10. P. A. Lachenbrach and R. M. Micke, Estimation of Error Rates in Discriminant Analysis, *Technometrics*, **10**, 1 (1968).
11. G. P. McCabe, Computations for Variable Selection in Discriminant Analysis, *Technometrics*, **17**, 103 (1975).

References

12. F. F. Blicke and R. H. Cox, *Medicinal Chemistry*, Vol. IV, Wiley-Interscience, New York, 1959.
13. M. Randić, On Characterization of Molecular Branching, *J. Am. Chem. Soc.*, **97**, 6609 (1975).
14. L. B. Kier, L. H. Hall, W. T. Murray, and M. Randić, Molecular Connectivity I: Relationship to Nonspecific Local Anesthesia, *J. Pharm. Sci.*, **64**, 1971 (1975).
15. L. H. Hall, L. B. Kier, and W. T. Murray, Molecular Connectivity II: Relationship to Water Solubility and Boiling Point, *J. Pharm. Sci.*, **64**, 1974 (1975).
16. W. T. Murray, L. H. Hall, and L. B. Kier, Molecular Connectivity III: Relationship to Partition Coefficients, *J. Pharm. Sci.*, **64**, 1978 (1975).
17. W. T. Murray, L. B. Kier, and L. H. Hall, Molecular Connectivity 6. Examination of the Parabolic Relationship between Molecular Connectivity and Biological Activity, *J. Med. Chem.*, **19**, 573 (1976).
18. L. B. Kier and L. H. Hall, *Molecular Connectivity in Chemistry and Drug Research*, Academic, New York, 1976.
19. A. J. Stuper and P. C. Jurs, Reliability of Nonparametric Linear Classifiers, *J. Chem. Inf. Comp. Sci.*, **16**, 238 (1976).
20. E. E. Swanson and W. E. Fry, The Pharmacological Relationship of Isometric Barbituric Acid Derivatives, *J. Am. Pharm. Assoc.*, **29**, 509 (1940).
21. C. Hansch and S. M. Anderson, The Structure–Activity Relationship in Barbiturates and Its Similarity to That in Other Narcotics, *J. Med. Chem.*, **10**, 745 (1967).

CHAPTER 7

Structure-Activity Studies of Olfactory Stimulants

Since the five fundamental senses are the only means by which an individual receives impressions from the surrounding environment, it is not surprising that a large amount of scientific research has been devoted to the understanding of these senses. The three physical senses—sight, hearing, and touch—have received a majority of the research effort in this field of sensory function and are now understood. However, the same cannot be said about the chemical senses—taste and smell. In the past, questions concerning these chemical senses were primarily pursued on the theoretical level with limited experimental work. However, the trend has changed and the fields of olfaction and taste have become active areas of research. While both these senses are interesting topics for research, this chapter deals with the structure–activity studies performed on odorant compounds.

Although olfaction is a complex process, it is generally accepted that the perception of odors involves the following steps: (1) the interaction of molecules from a volatile substance with receptors in the olfactory epithelium, (2) the transmission of nerve impulses in the olfactory bulb, (3) the processing of the impulses by the olfactory bulb, and (4) the delivery of olfactory information to the higher centers of the brain where the information is recognized and a response is emitted. To determine the details of these steps, research in several disciplines is necessary. Chemists must determine the chemical composition of odorous substances, as well as those molecular properties that are important for olfaction. Molecular biologists are needed to study the interaction of molecules and receptor cells, while physiologists and neurologists have the task of unraveling the neural activity at the olfactory bulb and the higher brain centers. Only with the collaborative efforts of these scientific disciplines will the olfactory process be determined. While there exists a plethora of unanswered questions, this chapter focuses on the molecular conformation of molecules and their perceived odor quality.

Although it appears that the source of chemoreceptory discrimination lies in the molecular structure of the odorants, the question, "What molecular properties govern the odor quality of an odorous substance?" still remains unanswered. While this problem has received a lot of theoretical speculation, only a limited amount of empirical research has been done.

One approach to this problem has been to assemble sets of odorants that have similar odors and then look for structural similarities. During the past few decades, this method has been used for small data sets and simple molecular properties. Unfortunately, none of the studies have been able to account for the odor qualities of a large and diverse collection of olfactory stimuli. Nevertheless, the possibility exists that a collection of several different molecular parameters can be combined to account for odor quality.

In this chapter two major structure–activity relation studies of olfactory stimuli are presented. The first deals with odorants that have a musk odor and the second with stimulants that affect the trigeminal nerve in the nasal cavity. However, before going into these two studies, we present some background information on olfaction to provide a common base for later discussions.

BASIC ANATOMY AND PHYSIOLOGY OF THE NOSE

The human nose is a bifunctional organ used for both breathing and detecting odors. Structurally, it is composed of the external nose and the nasal cavity. Although the external nose is a prominent facial feature, its primary function is merely to provide an opening into the nasal cavity, which contains the respiratory and olfactory regions. The majority of the nasal cavity, including three scroll-shaped turbinate bones, bears respiratory epithelium, which cleans, warms, and humidifies the incoming air (1). During normal respiration, most of the air stream is directed toward the posterior portion of the nasal cavity and down into the lungs. However, a small volume of air circulates up into the olfactory region, which is located high and to the back of the nasal cavity. Whenever an odor is noticed, more air can be sniffed into this region to facilitate the identification of the odor.

The olfactory region in humans consists of two patches of yellow tissue, one on each side of the nasal cavity, with each covering about 1 square inch. The olfactory epithelium is composed of both sensory olfactory cells and supporting epithelial cells. The entire epithelium is bathed in a thin layer of watery mucus supplied by Bowman's glands. The yellow pigment, which suggests a carotenoid, is found primarily in the supporting cells. Although this pigment may play a role in the olfactory mechanism, no data have been collected in favor of any specific activity (2).

The olfactory cells are long, narrow, bipolar neurons scattered throughout the olfactory epithelium and are responsible for the sense of smell. The cells are flasklike in shape and their lengths are directly proportional to the thickness of the epithelium. The olfactory vesicles emerge free at the mucus–cell interface and are provided with olfactory cilia or hairs. These hairs project from the tissue surface into the mucus layer covering them. Thus odorant molecules must dissolve, to some extent, in the watery mucus before reaching the olfactory hairs. Although an obvious role of these cilia is to increase the bare receptor surface area, the question of whether they are essential for olfaction is still under investigation (3).

The inner processes of the olfactory cells are very fine, unmyelinated nerve fibers. Each olfactory cell has one nerve fiber that maintains its individuality until it reaches the olfactory bulb of the brain. Together these nerve fibers constitute the olfactory or first cranial nerve (CN I). Each axon has a diameter of 0.2 to 0.3 microns, which makes it one of the smallest nerve fibers in the human body; consequently, it is extremely difficult to obtain electrophysiological measurements from this nerve. The olfactory nerve is solitary, having no other sensory or efferent fibers associated with it (3).

Free nerve endings from the trigeminal or fifth cranial nerve (CN V) are also found within the nasal cavity. In general the trigeminal nerve serves a protective function in that it responds to chemical irritants, such as ammonia or acid fumes, by influencing the secretion of mucus, the patterning of respiration, and the engorgement of the internasal erectile tissue (4). Since a later section in this chapter deals with trigeminally active compounds, further discussion on the role of this nerve in olfaction appears there.

In addition to the olfactory epithelium, a topographically distinct sensory area, called the Jacobson's organ or vomeronasal epithelium, can also be found in lower animals and infants; however, it is vestigial in adults. In the lower animals, this organ is connected to the lower portion of the nasal cavity through a narrow duct and contains receptor neurons whose axons terminate in the accessory olfactory bulb. The vomeronasal sensory cells are very similar to the olfactory sensory cell except they lack cilia (4). Although the vomeronasal organ does not have a role in adult olfaction, insight into the function of this organ would nevertheless aid in the overall understanding of the olfactory process.

Physiological investigations constitute a large percentage of the present activity in olfactory research. Progress in this area has been slow with very few major breakthroughs. The anatomy of the olfactory apparatus is known, but the knowledge has not increased the understanding of the olfactory process. Electrophysiological recordings of the olfactory nerve have been made (4–6), but the interpretation of the data has not yielded the solution to the olfactory code. However, the future may produce several of the answers to the questions currently being posed.

THEORIES OF OLFACTION

Over 2000 years ago, the poet Lucretius wrote (7), "You may readily infer that such substances as agreeably titillate the sense (of smell) are composed of smooth round atoms. Those that seem bitter and harsh are more tightly compacted of hooked particles and accordingly tear their way into our sense and rend our bodies by their inroads." Since then, theories on olfaction have abounded, with the largest number of new theories appearing during the last century.

The theories that are presently considered obsolete can be categorized as either wave or contact theories. Wave theories attribute olfaction to radiation emitted by odorant molecules, just as visible light affects the eye. Theories of this type no longer receive any attention, since it has been proven that olfaction occurs only with contact between the molecules and the olfactory region. Contact theories assume contact between the odorant molecule and olfactory receptors, but chemical reactions and intramolecular vibrations are considered to be responsible for the stimulation of the receptor. Although theories of this type were attractive in the early 1900s, sufficient data became available in the 1930s to disprove them (8).

The only theories presently being scrutinized by the scientific community are "whole-molecule effect" theories. In these odor is attributed to properties of the entire molecule rather than to functional groups or single molecular properties as is done in the contact theories. Vibrational, stereochemical, and molecular profile are all properties that have been incorporated into theories in this category. Since these theories are active areas of debate, they are discussed in more detail.

The vibrational theory of olfaction originated in 1937 when Dyson proposed that molecules having molecular vibrations in the 1400 to 3500 cm^{-1} range should be odorants (9). This theory attracted much interest at the time, but was quickly discarded because correlations between the infrared spectra of substances and their odors could not be found. In 1954 Wright revived the vibrational theory with one major modification: the active frequency range was lowered to the far infrared region (i.e., 50 to 500 cm^{-1}) where vibration of the whole molecule takes place (10,11).

In recent years, Wright's theory has been tested experimentally with the use of far infrared spectrometers. Unfortunately, there are few data to support this theory, but a large number of data to disprove it. Although one study of

47 musk odorants and 109 nonmusk compounds yielded statistically significant correlations (12), the fact that deuteration of an odorant molecule (which changes the far infrared absorption spectra) does not change its odor cannot be overlooked. Despite strong criticism on both experimental and theoretical grounds, this theory has not been completely discarded.

The stereochemical theory of olfaction, first outlined by Moncrieff (8) and later refined by Amoore et al. (13,14), attributes odor quality to the overall size and shape of the molecules. The theory is based on the "lock-and-key" concept so familiar in drug and enzyme theory. In this theory's early history, Moncrieff suggested that there were between 4 and 12 types of receptor sites, each corresponding to a primary odor. However, no experimental work was done to confirm the theory.

Amoore contributed considerably to this theory by first trying to determine the number of different receptor sites and then the dimensions of each site. These refinements were accomplished by an extensive literature search to find compounds that exhibited similar odor qualities, followed by the inspection of molecular models of the different odorants in an attempt to determine the receptor site shape and size. The result of his work initially yielded seven primary odor types: camphoraceous, pungent, ethereal, floral, pepperminty, musky, and putrid. However, these odor classes were not found to depend on size and shape to the same degree. The ethereal, camphoraceous, and musky odors depended mainly on size; whereas the minty and floral classes were more contingent on the molecular shape. The remaining two groups, pungent and putrid, were found to be dependent on the electronic nature of the molecule rather than on either shape or size. The fact that two of Amoore's primary odors do not depend on the shape and size of the molecule indicates that there may be other factors important in the olfactory process. Nevertheless, one of the better rationalizations of our knowledge on olfaction is offered by this theory.

The final theory to be discussed is the molecular profile theory proposed by Beets in 1957 (15) in which both functional group location and molecular structure are used to explain odor quality. According to this theory, the functional group is responsible for the orientation of the molecule at the receptor, while the profile of the molecule presented to the receptor determines the odor quality. Although studies have shown this observation to be true in some cases and have led to the development of some new odorants, the theory has not been refined sufficiently to determine rules for functional group placement nor the shapes of important profiles. Nevertheless, this molecular profile theory is still important for explaining the odor qualities of some groups of odorants.

In these different theories, there appear to be two underlying properties that are necessary for a substance to be odorous. First, the material must be of sufficient volatility so that it can travel to the olfactory region in the nose, and second, it must have lipid solubility as well as some water solubility so that it can reach the olfactory receptors. Beyond these two properties, there is very little agreement as to which molecular properties are responsible for the many different odors known to mankind. This is where techniques for doing structure–activity studies can be of most value.

The goal of any study of structure–activity relation (SAR) is to correlate biological activity of a compound with its molecular structure. During the procedure, rules are obtained to assist in the development of new compounds that will have the desired biological activity. Although SAR have long intrigued the minds of chemists, it has only been within the last century that sufficiently large data sets have become available to permit such studies.

Since knowledge of the molecule–receptor mechanism is not a prerequisite for doing a SAR study, the method is well suited for investigating a large number of different systems, including olfaction. In fact, most theories of olfaction have been attempts at describing structure–activity relations.

The early efforts at SAR studies in olfaction were mainly attempts to correlate a compound's odor quality to a single molecular property using simple linear regression analysis. Although statistically significant results were obtained in isolated cases, the ability to predict the odor quality of a large data set was never demonstrated. Even an attempt to combine 25 physicochemical parameters to predict an odor space by employing multidimensional scaling techniques (16) produced only fair results.

The Hansch approach has been widely accepted in pharmacology as a versatile means of understanding SAR in drug systems; however, its utility in olfactory research is somewhat restricted. The Hansch approach was used by Boelens (17) to investigate some bitter almond and musk odorants. The reported work was done on data sets of 16 odorants per study, with the most significant parameter found being the 1-octanol/water partition coefficient in each case. The results suggested that the odorants' activities (in this case the relative quality of the odorant as compared to a standard by a panel of perfumers) were due primarily to the compounds' solubilities alone. The structural features important for the musk or bitter almond odors were not discussed. Therefore, until accurate methods are developed to obtain reproducible quantitative olfactory data, the Hansch approach will have only limited applicability in olfactory research.

Although a large number of qualitative data exist for many compounds, SAR studies have only been done on limited data sets using only simple correlation techniques with two or three variables in any one study (18,19). One factor limiting the investigations in qualitative SAR studies has been the lack of good techniques for handling data of this type. However, pattern recognition techniques are well suited for handling qualitative data as well as quantitative data.

Pattern recognition techniques are uniquely suited for doing qualitative SAR studies because of various characteristics of the procedures. First, the techniques attempt to provide definitions of similarity between diverse groups of data. They are able to deal with high dimensional data where more than three measurements are used to describe each object or event. Furthermore, pattern recognition techniques can handle data in which the relationships are discontinuous, as well as multisource data where each measurement can be the result of an independent experiment. This attribute is very important, since SAR studies involve data of this type. Finally, techniques are available for selecting important features from a larger set of measurements, thus allowing studies to be done on systems where the exact relationships are not fully understood.

Therefore, the major problem in applying pattern recognition to SAR studies is determining which molecular properties to use for describing a set of compounds that have the desired activity. Naturally, the three parameters employed in the Hansch approach could be used with pattern recognition techniques, but determining the values for these parameters for a structurally diverse data set is extremely difficult, to say the least. However, since the structure of a molecule completely defines its physical properties and, consequently, its biological activity, it should be possible to decompose the structure into a set of descriptors that are sufficient to place the compound in the class in which it belongs.

ANALYSIS OF MUSK ODORANTS

The class of compounds commonly known as musks was chosen for the initial study primarily because it has a highly characteristic odor quality that is rarely confused with other odors. Therefore, a data set composed of this class of odorants is relatively free of misclassified compounds. Such a well-characterized data set is important for providing a fair test of the capabilities of pattern recognition techniques for doing SAR studies in olfaction. However, this factor was not the only one to play a role in the selection of this class of compounds.

Since the 1950s, perfume manufacturers have done considerable research in developing synthetic musk odorants to replace the diminishing supply of natural musk compounds. Consequently, a large amount of information is available on musk odorants (15, 18, 20). This is not true for any other major class of odorants. Furthermore, this class of compounds is structurally interesting since it contains a variety of different structural types, including some

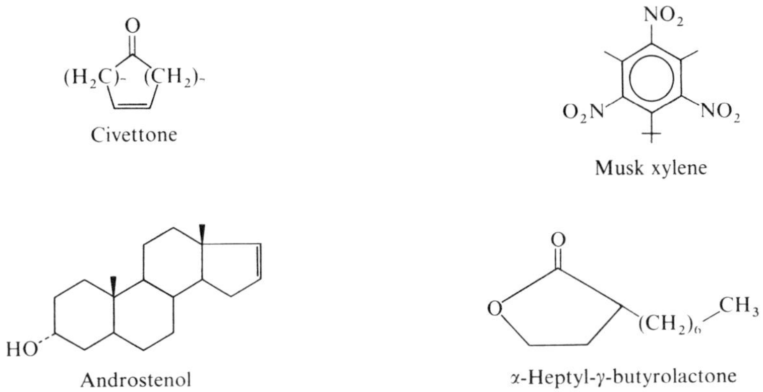

Analysis of Musk Odorants

Figure 7.1 Example structures of musk odorants.

steroids. In general the SAR study of musk odorants using pattern recognition presented itself as a challenging problem with a high probability of success.

For this study of musk compounds, a data set of 300 unique compounds was selected from the library of odorants compiled by Amoore (21). Sixty of these compounds were musk odorants; they included 23 macrocyclics, 19 polynitrobenzenes, 11 steroids, 5 γ-butyrolactones, and 2 compounds having other structural types (see Figure 7.1 for examples of each structural type). Although 16 of these musk compounds were classified by Amoore as weak musks or as having other odor overtones (see reference 21), enough strong musks were present to assure a good representation of musk odorants (see the Appendix for a complete listing of the musk odorants used in this study).

The 240 compounds in the nonmusk class were randomly selected from the remaining 6 odor quality classes. The data sets included 49 camphor, 44 floral, 32 ethereal, 41 mint, 51 pungent, and 23 putrid compounds. In this group a large number of different functional groups and structural types were present, thus assuring a good representation of nonmusk odorants.

After the structures were entered into data files, a series of descriptors was generated for each compound. Fragment descriptors were included to provide some general information about the compounds' chemical natures. The total number of atoms, bonds, carbon atoms, oxygen atoms, nitrogen atoms, single bonds, double bonds, triple bonds, aromatic bonds, and a weighted summation of the four basic bond types were all generated for this data set. The five fragment descriptors excluded from this list contained very little information, since only a few compounds had nonzero values.

To obtain some information about the chemical functionality and structural makeup of the compounds, a series of substructure descriptors was generated. A screening of the 46 substructures given in Table 7.1 indicated that only 41 of them had values greater than zero for at least 10% of the data set members. Since 10 of the substructures were sought in the data set using both the general search and the specific search, a total of 51 substructure descriptors was obtained.

The seven geometric descriptors described previously were then generated for this data set using the three-dimensional coordinates obtained from MOLMEC. It was felt that the overall shape of the molecules, as reflected by these geometric descriptors, would be important for the separation of the musk compounds from the nonmusks, since Amoore obtained good correlations for this odor quality. The generation of environment and molecular connectivity descriptors was postponed until the results of using these 68 descriptors were obtained.

Table 7.1 Substructure Descriptors[a]

1. (S) —C—*
2. (S) —C—*
3. (S) *C—C—
4. (S) *C—C—C—
5. (S) —C—* (=)
6. (B) —C—
7. (B) —C—
8. (B) —O—
9. (B) —C—
10. (B) C—C—
11. (B) C—C—
12. (B) —C—C—
13. (B) —C=
14. (B) —C=O
15. (B) C—C—C
16. (G) —C
17. (G) —O
18. (G) =O
19. (G) C—C—
20. (G) ═C═
21. (G) ═C═
22. (G) ═C═C═
23. (G) ═C═C═
24. (G) ═C═C═
25. (G) ═C═C═C═
26. (G) ═C═C═C═C═
27. (G) C—C—

Analysis of Musk Odorants

Table 7.1 (*Continued*)

28. (G) ⋕C—C⋕	37. (G) —C=
29. (G) C—C—C—	38. (G) —C—O—
30. (G) —C—C— 　　　　｜	39. (G) —C—C—C—
31. (G) —C—C— 　　　　｜	40. (G) —C—C—C— 　　　　　　｜
32. (G) —C—C— 　　　｜　｜	41. (G) C—C—C— 　　　　｜
33. (G) —C—C— 　　　｜｜　｜	42. (G) C—C—C— 　　　　｜
34. (G) C—C= 　　　｜	43. (G) C—C—C— 　　　　｜｜
35. (G) —C—C=	44. (G) —C—C—C— 　　　　　　｜
36. (G) —C—C= 　　　｜｜	45. (G) —O—C=O
	46. (G) —C—C—C—C—

a (⋕): Aromatic bond type; (∗): atom is specified to be in a six membered ring; (S): specific search was done; (G): general search was done; (B): two searches were done, one specific and one general.

The 68 descriptors generated for each compound in the previous steps were subsequently combined into pattern vectors. The vectors' components were then preprocessed so that each component had a mean of zero and a standard deviation of unity. All components were then multiplied by a constant multiplier of 100 to prevent the loss of information due to truncation.

The initial test was to determine if a linear discriminant function, $f(\mathbf{X}_i)$, could be found such that $f(\mathbf{X}_i) > 0$ for $\mathbf{X}_i$ taken from the musk class and $f(\mathbf{X}_i) \leq 0$ for all other compounds. Using the training procedure described previously, a decision surface that correctly classified the entire data set was found. Based on this knowledge, several studies were conducted to determine which of the 68 available descriptors were most important for the separation.

Instead of using the feedback feature selection on the 68 component pattern vectors, it was decided to first test the geometric, fragment, and substructure descriptors individually with respect to their ability to separate the data set. It was found that neither the fragment nor the geometric descriptors were able to produce a linear discriminant function. Even the combination of these

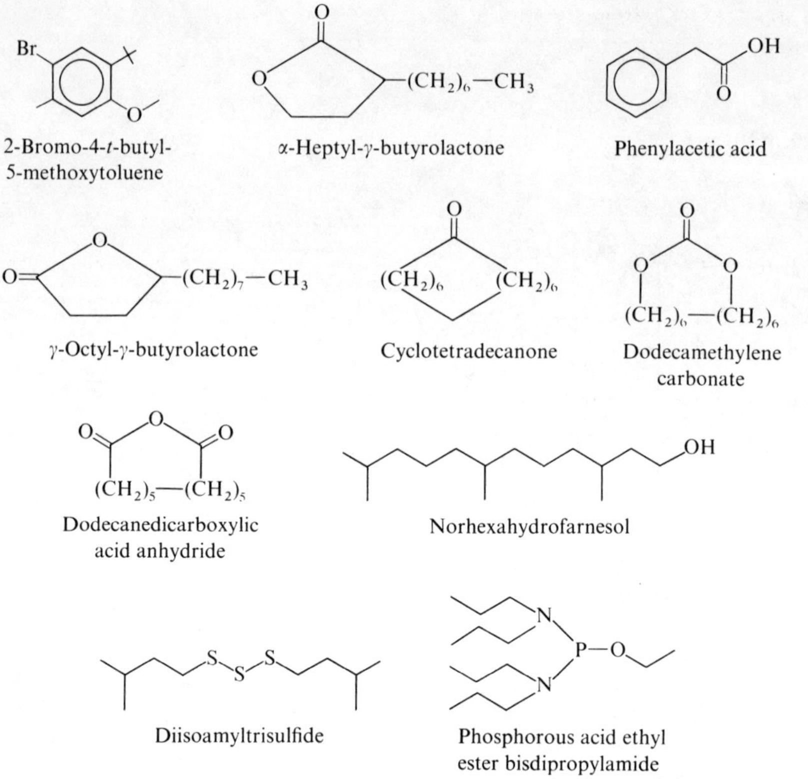

Figure 7.2 Compounds misclassified using only geometric descriptors.

descriptors was unsuccessful in finding a linear surface. However, it was found that only a few compounds were preventing linear separability. When the 7 geometric descriptors were used alone, only the 10 compounds shown in Figure 7.2 were misclassified. Upon the inclusion of the 10 fragment descriptors in each pattern vector, only the top 3 compounds in Figure 7.2 prevented linear separability. Since these three compounds have been characterized as weak musk odorants, it was not surprising that they were confusing the learning machine, and thus preventing linear separability. Instead of removing these compounds from the data set and then feature selecting, it was decided to test more descriptors on the entire data set.

When the 51 substructure descriptors were tested, a linear disriminant function was found that correctly classified the entire data set. To provide a measure of the predictive ability of these descriptors, the data set was subdivided into a series of training and prediction sets. Eighty unique sets were randomly

Table 7.2 The Distribution of the Musk Data Set Members into Training and Prediction Sets

Group	Training Sets Musks	Training Sets Nonmusks	Prediction Sets Musks	Prediction Sets Nonmusks	Number Sets Generated
A	50	200	10	40	20
B	48	192	12	48	20
C	45	180	15	60	20
D	42	168	18	72	20

generated. The number of compounds included in these sets from each odor class is given in Table 7.2.

Each training set was used to develop a decision surface, which was then used to classify the members of its corresponding prediction set. By using the 20 randomly selected sets in each group, an average predictive ability can be calculated. Table 7.3 contains the results of these predictive ability studies using all 51 substructure descriptors.

The training set that gave the highest predictive ability in each group was subsequently used for feature selection. The number of substructure descriptors remaining after feature selection and their average predictive abilities for each group's training sets are also given in Table 7.3. In each case at least two-thirds of the initial 51 descriptors were removed during feature selection, while the average predictive ability increased slightly. In all cases the entire data set was linearly separable with the reduced number of descriptors. The substructure descriptors retained during feature selection and the groups in which each substructure were found to be important are given in Table 7.4.

Table 7.3 Predictive Ability Studies Using Only the Substructure Descriptors

Group	Initial Predictive Ability[a]	Final Predictive Ability	Final Number of Descriptors
A	95.1	96.2	14
B	94.6	95.3	15
C	94.6	95.8	16
D	94.6	95.1	14

[a] Fifty one substructure descriptors used in each pattern vector.

Table 7.4 The Substructure Descriptors Retained during Feature Selection

Substructure	Search Type[a]	A	B	C	D	Substructure	Search Type[a]	A	B	C	D
1. —C	S	*	*	*	*	15. —C—C—C—C—	G	*		*	
2. C⩭	G	*	*	*	*	16. C—C(—)—C	G			*	*
3. —C—	G	*	*	*	*						
4. —C—	S	*	*	*	*	17. —C—O—	G		*		*
5. —O—	S	*	*	*	*	18. ⩭C(—)⩭C(—)⩭	G			*	
6. —O	S		*	*	*	19. —C—C—	S	*			
7. ⩭C⩭C	G	*	*	*		20. —C(=)—	G			*	
8. ⟩C—C⟨	G	*	*		*	21. —C—C—C—	G	*			
						22. —C—C(=)—	G				*
9. —C(=)—	S	*		*	*	23. —O—	G	*			
10. —C(—)—	S		*	*	*	24. C—C(—)—C	S	*			
11. —C—C	S	*		*		25. —C(=)—C	G			*	
12. ⩭C(—)⩭C⩭	G	*	*			26. —C(—)—C—	G			*	
13. —C(—)—	G			*	*						
14. —C(—)—C	G	*		*		27. —C—C(—)—C	G				*

[a] G = general search, S = specific search, (⩭) is an aromatic bond type.

As can be seen, there is considerable overlap of descriptors retained from group to group with over one-third of these 27 descriptors being retained as important features in at least three-fourths of the groups. Since 24 of the 51 substructures were always excluded during feature selection, it was assumed that they were not essential for the separation of this data set. Therefore, they were not employed in any of the subsequent studies.

Table 7.5 The Predictive Ability Studies Using the 44 Combined Descriptors

Group	Initial Predictive Ability[a]	Final Predictive Ability	Final Number of Descriptors	Final 13 Descriptors[b]	Final 11 Descriptors[c]
A	95.3	96.6	16	97.5	96.4
B	96.1	95.9	15	97.6	96.8
C	95.6	96.4	15	96.8	96.6
D	95.7	97.1	14	97.8	96.7

[a] The 27 substructure descriptors plus the 7 geometric and the 10 fragment descriptors were included in each pattern vector.
[b] All 13 descriptors given in Table 7.7 were used in each pattern vector.
[c] All descriptors in Table 7.7 except for numbers 2 and 5 were included in each pattern vector.

The 27 substructures found to be useful in the foregoing study were then combined with the 10 fragment and 7 geometric descriptors to form pattern vectors containing 44 descriptors each. As in the substructure study, the 80 training and prediction sets were used to obtain an average predictive ability for each group. Feature selection was then carried out using the best training set in each group. The results of these predictive ability studies are given in Table 7.5. In each group about two-thirds of the initial 44 descriptors could be removed while a high predictive ability was maintained. Table 7.6 contains the descriptors retained during feature selection. (The substructure numbers in this table correspond to the index numbers used in Table 7.4.) As can be seen, some of the substructure descriptors found to be useful in the previous study were replaced by the fragment and geometric descriptors with an increase of the predictive ability for each group resulting from these exchanges (compare results in Tables 7.3 and 7.5).

Although the descriptors found to be useful for each group could have been used to predict the class designation of unknown odorants, a better method is to develop a classifier on the basis of the entire data set. When feature selection was performed on all 300 compounds, the 13 descriptors given in Table 7.7 remained out of the initial 44 descriptors. Although linear separability was still maintained when descriptors 2 and 5 in Table 7.7 were removed, the predictive ability decreased for each group, indicating a removal of some information (see last two columns in Table 7.5). Included in Table 7.7 are the predictive abilities for each individual descriptor to classify the entire data set. These percentages should be compared to 80%, which would be obtained by classifying the entire data set as nonmusks. The best single predictor in this

Table 7.6 The Descriptors Retained during Feature Selection of the 44 Combined Descriptors

Descriptor[a]	Included in Groups			
	A	B	C	D
1. Number of oxygen atoms	*	*	*	*
2. Substructure number 3	*	*	*	*
3. Substructure number 9	*	*	*	*
4. Substructure number 5	*	*	*	*
5. X radius of gyration	*	*	*	*
6. Number of single bonds	*	*		*
7. Number of double bonds			*	*
8. Substructure number 1	*		*	*
9. Substructure number 8	*	*		*
10. Substructure number 15	*		*	*
11. Substructure number 27		*	*	*
12. Number of aromatic bonds		*		*
13. Substructure number 6		*	*	
14. Substructure number 18	*		*	
15. Substructure number 13		*	*	
16. Substructure number 23	*	*		
17. Y radius of gyration	*		*	
18. Number of carbon atoms			*	
19. Substructure number 2	*			
20. Substructure number 12	*			
21. Substructure number 21				*
22. Substructure number 26	*			
23. Substructure number 25	*			
24. Substructure number 16				*
25. Substructure number 17		*		
26. X radius/Z radius ratio		*		
27. Y radius/Z radius ratio			*	

[a] The substructure numbers used in this table correspond to the index numbers in Table 7.4.

Table 7.7 The 13 Descriptors Remaining after Feature Selection Using the Entire Data Set [a]

	Classification Percent Correct	Descriptor
1.	84.3	Total number of oxygen atoms per molecule
2.	82.3	Total number of double bonds per molecule
3.	80.0	Total number of aromatic bonds per molecule
4.	86.7	Longest principal radius
5.	80.0	Shortest principal radius
6.	80.0	C—C(—C)—C (branched)
7	80.0	C—C(—C)—C
8.	86.0	≡C≡C≡ (aromatic C—C)
9.	90.3	—C—
10.	80.7	—C
11.	80.0	—C(=)—
12.	80.0	—O—
13.	83.0	—C(—)—

[a] (≡): Aromatic bond.

list of descriptors is the methylene substructure (descriptor 9, Table 7.7), which reflects the fact that macrocyclic musks contain a large number of these substructural units.

To further test the predictive ability of the descriptors given in Table 7.7 and the discriminant function associated with them, nine previously unused musk odorants were used to test the classifier. The odorants, shown in Figure 7.3 were entered into the ADAPT system and pattern vectors incorporating only the best 13 descriptors were generated for each compound. After preprocessing, the nine unknowns were classified as musk or nonmusk using the weight vector trained on the entire data set of 300 odorants. All nine compounds were correctly classified as musk compounds. The correct classification of the five nitro musks and the one macrocyclic musk in the data set of unknowns was expected, since the training set contained structurally similar

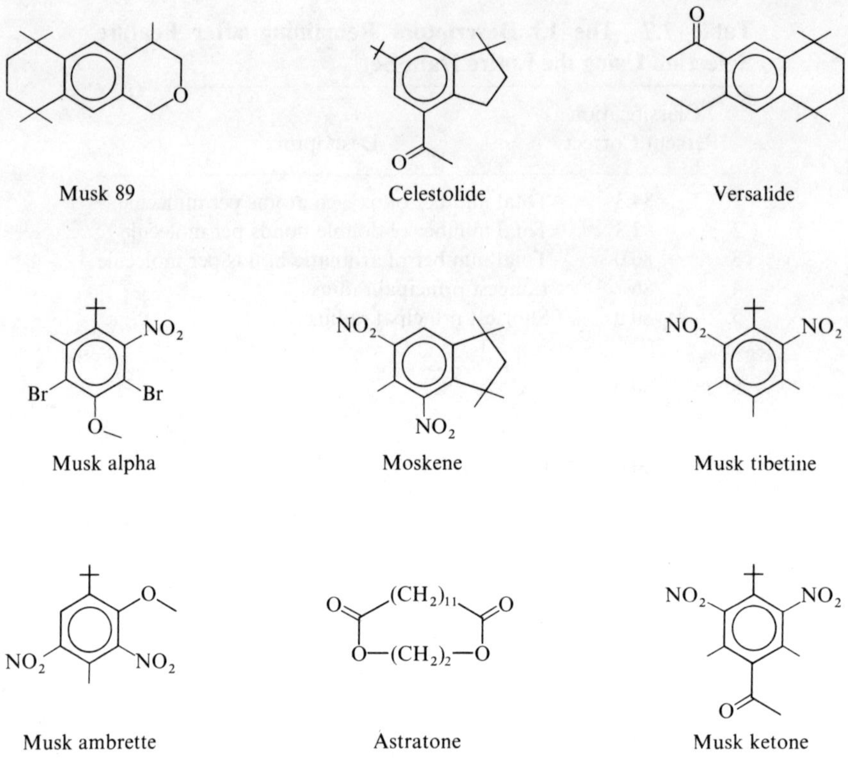

Figure 7.3 Structural diagrams of the nine unknown musks.

compounds. However, the correct prediction of the remaining three unknowns was interesting, since these were new structural types never employed in the training of the discriminant function. Thus the classifier was able to recognize new categories of musk odorants on the basis of a few molecular parameters that were derived from musk odorants of different structural types. Therefore, it appears that these parameters reflect the molecular properties common to musk odorants.

A Common Structural Unit For Musk Odorants

The pronounced structural similarity between steroid and macrocyclic musk compounds has been known for a number of years. This similarity can be seen in Figure 7.4, which uses Civettone (a macrocyclic musk) and Androstenol (a strong steroid musk) as examples. The fact that both of these compounds have the same number of peripheral atoms is very interesting, especially since other

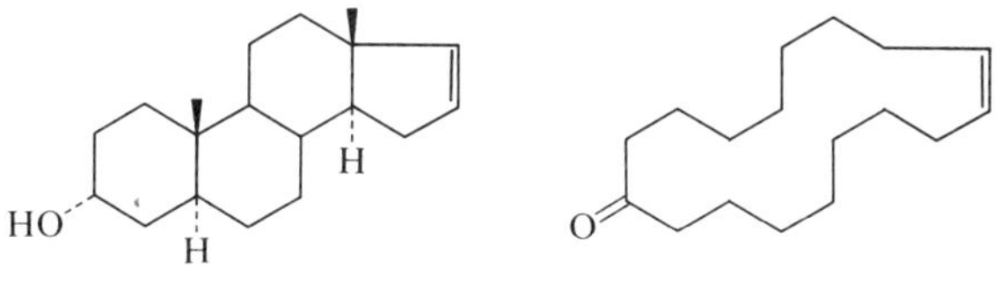

Analysis of Musk Odorants

Androstenol   Civettone

Figure 7.4 A structural comparison of a steroid and a macrocyclic musk odorant.

macrocyclic ketones, lactones, and carbonates having 15 to 17 ring atoms are also perceived as musks (8). The odor similarity between these two classes of compounds can best be explained by the ability of the macrocyclic compounds to assume a structural conformation similar to that of the rigid steroid structure. However, the musk odor of polynitrobenzene musks, as well as that of other synthetic benzene musks, cannot be explained by this peripheral atom model.

A casual comparison between polynitrobenzene and steroid musks, would indicate that there is no structural similarity other than simple fragments such as methyl groups or hydroxide groups. However, a more careful examintion of these two classes of compounds revealed the common structural unit shown in Figure 7.5. As can be seen, the two substructures are identical except for three carbon to carbon bonds. When stick models of these two substructural units were constructed and compared, it was found that the three-dimensional spacial relationships of atom attachments A, B, and C varied by only a small amount. In checking the 11 steroid and 19 polynitrobenzene compounds among the musk data set, only 3 nitrobenzene musks were found that did not contain this structural unit.

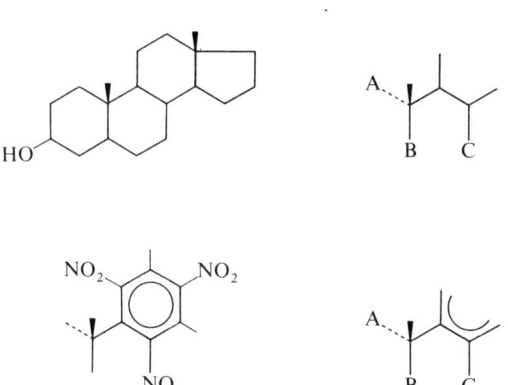

Figure 7.5 The common structural unit for steroid and polynitrobenzene musk odorants.

Table 7.8 The Nine Best Descriptors[a]

1. The number of oxygen atoms per molecule
2. The number of aromatic bonds per molecule
3. X radius of gyration
4. Z radius of gyration
5. —C—
6. —O—
7. —C(=O)—
8. C—C—C≃C—
9. C—C—C—C—

[a] (≃): Aromatic bond.

According to this information, these two substructures should be useful for the separation of musk odorants from nonmusks. To test this idea, these two substructures were collated with the 13 descriptors given in Table 7.7 and were then employed to describe the data set used in the previous study. After variance feature selection using the entire data set, only the 9 descriptors shown in Table 7.8 remained out of the original 15 descriptors. The two new substructures were not removed during feature selection. In Table 7.9 the

Table 7.9 A Comparison of the Predictive Abilities of Different Descriptor Sets

Prediction Set	15 Descriptors[a]	13 Descriptors[b]	9 Descriptors[c]
A	97.6	97.5	97.8
B	97.7	97.6	97.9
C	96.7	96.8	97.3
D	97.2	97.8	97.2

[a] Descriptors in Table 7.7 plus the two substructures shown in Figure 7.5.
[b] The descriptors shown in Table 7.7 alone.
[c] The descriptors shown in Table 7.8.

predictive abilities for these 9 descriptors are compared with the predictive abilities for the previous "best" set of 13 descriptors shown in Table 7.7. As can be seen, there was a slight improvement for 3 out of the 4 groups, thus indicating no loss of information by going down to 9 descriptors.

To further test the predictive power of these descriptors, a prediction data set, composed of previously unused musk and nonmusk odorants, was generated. For this data set 120 nonmusk compounds were selected from the master library of compounds originally obtained from Amoore's data set (21). None of these 120 nonmusks were ever used in any of the previous studies. The camphor, floral, minty, and pungent odor qualities were all represented in this category. For the musk category, 121 new musk compounds were obtained from an article by T. F. Wood (20). While 31 of these musk odorants were polynitrobenzene, the other 90 compounds were bicyclic and tricyclic aromatic musks that were never incorporated into the data set used to select the descriptors employed in this study. After the structures were entered into the computer system and their three-dimensional models were obtained, the 15 descriptors necessary for this study were generated.

The classification of these 241 compounds into musk/nonmusk categories was done using two weight vectors trained with the 300 member data set described in the beginning of this section. The first weight vector utilized the 13 descriptors in Table 7.7 as its components, while the second weight vector used the 9 descriptors shown in Table 7.8. The classification results for these two vectors are given in Table 7.10.

For the classification of the nonmusk odorants, the two weight vectors performed identically in that each one missed the same three compounds, the structures of which are shown in Figure 7.6. A comparison of these structures with the descriptors given in Tables 7.7 and 7.8, readily shows that these three odorants do contain the structural properties found to be important for the classification of musks. The most obvious misclassification is the macrocyclic compound that contains 12 ring atoms instead of the 15 to 17 ring atoms that most macrocyclic musks have. The cause of these misclassifications is most

Table 7.10 Prediction Results Using Two Weight Vectors Trained for Classifying Musk Odorants

Weight Vector	Correct Classifications			
	Musk Compounds		Nonmusk Compounds	
Nine components	111/121	91.7%	117/120	97.5%
Thirteen components	109/121	90.1%	117/120	97.5%

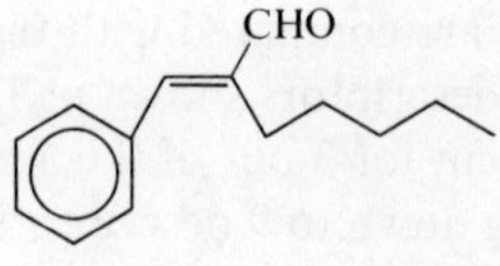

α-*n*-Amylcinnamaldehyde

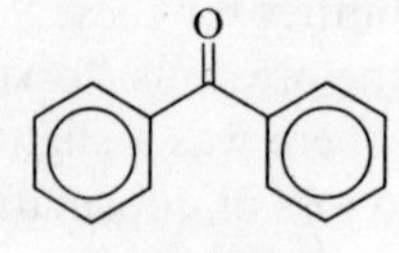

Benzophenone

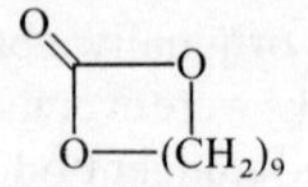

Nonamethylene carbonate

Figure 7.6 The three nonmusk compounds misclassified by the two weight vectors.

likely an inadequate training set of compounds rather than the descriptors employed. The only method to improve the predictive ability of these nonmusks would be to include more representative nonmusk odorants in the training set.

For the musk odorants, the number of misclassifications was almost identical for the two weight vectors. However, in all except three cases, a compound misclassified by one weight vector would be correctly classified by the other weight vector. In Figure 7.7 the three compounds missed by both weight vectors are shown. In all three cases the compounds do not contain the exact molecular descriptors used in developing the weight vectors, but rather they contain related structural features. For example, the two tricyclic musks do

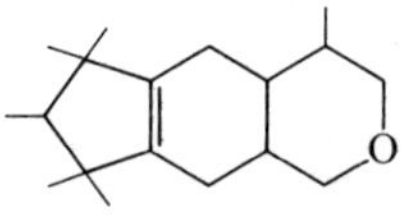

1a,4a,5,9-Tetrahydro galaxolide

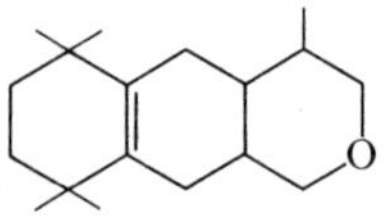

1a,4a,5,10-Tetrahydro musk 89

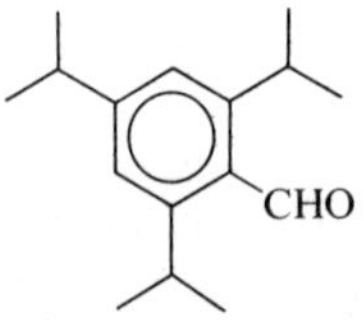

2,4,6-Triisopropylbenzaldehyde

Figure 7.7 The three musk compounds misclassified by both weight vectors.

contain the substructural unit shown in Figure 7.5, except a double bond replaces one of the single bonds. In the benzene musk, the substructural unit is almost present except for one methyl group. Although the human mind can recognize these structural similarities, the descriptors employed in these studies are not flexible enough to account for such small changes. However, it should be pointed out that these three compounds have been classified as weak musks with amber and woody odor notes (20).

In checking the 121 new musk odorants, it was found that only 10 of these compounds did not contain either of the substructures shown in Figure 7.5. These 10 compounds were the ones misclassified by the weight vector incorporating these two substructural descriptors (see Table 7.8). Furthermore, in checking the 435 nonmusk compounds in a master library file, only 5 steroid compounds, with a cedarwood odor, were found to contain either one of these substructures. Although these two substructures are not absolutely necessary for the musk odor, the fact that they are contained in 78 % of all the musk odorants used in these studies indicates that they are highly characteristic of this odor quality. In the future it would be interesting to study the odors of compounds that contain these substructural units.

These studies have shown that pattern recognition techniques can be used to extract important features from a collection of parameters for a given class of odorants, as well as to predict whether an unknown compound will have a musk odor. However, the successful results cannot be entirely attributed to the techniques employed. The choice of the data set and the descriptors were also important factors. Musk compounds were chosen for this initial investigation because of their highly characteristic odor quality. The results of these studies fully support the observation that musk odorants are easily distinguishable from other odorants.

The ability of the fragment and geometric descriptors alone to classify all but three weak musks indicates that information about the compound's structural shape, as well as its chemical composition, is necessary to separate musk odorants from nonmusks. Therefore, it is not surprising that the substructure descriptors alone were able to linearly separate the entire data set, since these descriptors encode both structural and chemical information. As would be expected, the best set of descriptors for predicting musk odorants was found to be a combination of all three types of descriptors (cf. Table 7.8). Although these nine descriptors were found to have the highest predictive ability and were able to correctly classify all but 10 unknown musks, they should not be considered to be the optimum set of descriptors for classifying musk odorants, since they were found through a heuristic search of a limited number of molecular descriptors. This point is exemplified by the fact that two substructural descriptors were able to replace six other descriptors in this study.

Although the actual role of these nine descriptors in the olfactory process is not clear, the fact that they fall into two categories (i.e., chemical composition and geometric shape) indicates that there may be two steps involved in producing the musk odor. One hypothesis is that the chemical composition reflects the ability of the compound to pass from the vapor, through the mucosa, to the site of interaction, and the geometric shape of the molecule determines how well it fits into a receptor site. However, more empirical research is needed on this topic before this conjecture can receive adequate verification.

In general, these studies have demonstrated that pattern recognition techniques are well suited for finding invariant properties among a large data set of compounds having the same odor quality. As for musk odorants, molecular shape seems to be an important factor for accurate prediction, but it is by no means the only factor. As shown above, a few molecular descriptors are capable of producing a good classifier for predicting musk odorants. The correct classification of 111 unknown musk compounds is a strong indication that the descriptors incorporated into the classifier reflect molecular properties that are common among musk odorants. The fact that new structural types were recognized in these prediction studies lends weight to the argument. Although musk odorants were the focus of this work, other odor qualities can be studied in a similar manner and are the topic of the next section.

ANALYSIS OF TRIGEMINALLY ACTIVE COMPOUNDS

Intranasal neural structures play an important role in monitoring the intake of airborne chemicals into the human respiratory system. As is mentioned in the previous section, two separate nerve systems are found within the nasal cavity. The olfactory nerve is responsible for detecting a wide range of concentrations of numerous vapors and transmitting information to higher brain centers, resulting in the perception of odor. Free nerve endings from the trigeminal nerve (CN V) are also distributed throughout the nasal mucosa and are responsible, in part, for the detection of irritating chemical vapors. However, unlike the olfactory nerve, the trigeminal nerve has efferent branches that can influence the secretion of mucus, the patterning of respiration, and the engorgement of intranasal erectile tissues, all of which affect the flow of air in both the upper and lower respiratory tracts (see reference 4 for a review). These responses are presumably protective, since prolonged exposure to some trigeminal stimulants can produce serious dysfunctions in humans (22–24). Under extreme conditions of air pollution, chemical stimulation of the trigeminal nerve can result in the reversal or even cessation of breathing itself (see references 4 and 25–29).

In a recent study (29), it has been determined that many volatile chemicals commonly used in olfactory research can be perceived by patients who exhibit no olfactory nerve function. Sensations such as stinging, burning, coolness, warmth, or pain are typically reported by these anosmics. Exactly which sensations occur depends on the chemical being presented and its concentration. The chemicals that can be detected by anosmic observers differ from those that cannot be detected with respect to several physicochemical parameters, including molecular weight, water and lipid solubility, boiling point, and dipole moment (29). However, it is not clear which parameters are responsible for producing the sensations in anosmics.

Since the trigeminal nerve branch in the nasal cavity is important for both detecting potentially hazardous vapors and interacting with the olfactory system in the perception of odorants (30), establishing the nature of the perceptual responses generated by this system, as well as the molecular properties responsible for the interaction, is important for the complete understanding of the sense of smell. It is also of particular importance to olfactory theorists who have sought to find "pure olfactory" stimulants (see reference 4). To gain information on this nerve system, quantitative data on the perceptual responses of humans to intranasal trigeminal stimulation by inhaled vapors was obtained from psychometric ratings of the perceived intensity, pleasantness, coolness, warmth, and presumptive safety (i.e., the observer's estimate as to the probable safety of the chemical if inhaled frequently) of 47 diverse chemicals. Three groups of observers were used: (1) anosmic observers lacking olfactory (CN I) nerve function, but not trigeminal (CN V) function; (2) normal observers instructed to pay attention to only the trigeminally mediated sensations within the nasal cavity ("trigeminal focus" group); and (3) normal observers instructed to rate the overall smell experience in the traditional fashion. Studies of structure–activity relations were then conducted to determine whether the perceived trigeminal intensities, as reported by the anosmics, could be predicted on the basis of a few physiochemical or molecular structural properties. In these studies, multiple regression analysis and discriminant function analysis were used in conjunction with pattern recognition techniques to test numerous parameters as to their ability to predict the psychometric measures of trigeminal intensity.

Procedure for the Trigeminal Study[1]

Observers for this study included 15 anosmic and 30 normal individuals who were divided into three experimental groups. The 15 anosmics included 7 males diagnosed as having a congenital disorder in which the olfactory bulbs

---

[1] The collection of the experimental data was performed at the Monell Chemical Senses Center under the direction of Dr. Richard L. Doty.

and stalks are aplastic (see references 31 and 32). Two male and 2 female anosmics reported never having been able to smell since childhood and exhibited no CN I function upon psychophysical examination. One male anosmic lost his sense of smell following an anterior craniotomy, while another became anosmic following the malfunction of his gas mask in a phosgene and tear gas exercise during combat training for World War II. One female anosmic lost her sense of smell in adulthood following a skull trauma, while another lost her sense of smell in adulthood after an influenza bout. These 15 individuals made up the anosmic group.

The 30 normal individuals were divided into two groups. The "trigeminal focus" normal group ($N = 15$) consisted of 7 males and 8 females. The normal experimental group ($N = 15$) consisted of 8 males and 7 females. All 30 individuals comprising these two groups exhibited normal olfactory acuity upon psychophysical examination (see reference 29).

The 47 compounds used as stimulants in this experiment were chosen so as not to exceed the moderate rating in Sax's system of toxic hazards regarding inhalation (33). These compounds were also chosen to exhibit a reasonably large variety of different chemical structures and to contain a homologous series of aliphatic acids (Table 7.11 contains the compounds investigated). All

Table 7.11 Detection and Intensity Data for Anosmics

Compound	Proportion Detecting	Intensity Mean	Std.
1. Decanoic acid	0/15	0.00	(0.00)
2. Vanillin	0/15	0.00	(0.00)
3. Phenyl ethyl alcohol	1/15	0.13	(0.50)
4. Eugenol	1/15	0.13	(0.50)
5. Coumarin	2/15	0.13	(0.34)
6. Nonane	3/15	0.27	(0.57)
7. Octane	3/15	0.27	(0.57)
8. Indole	3/15	0.53	(1.20)
9. α-Terpineol	5/15	0.53	(1.02)
10. Geraniol	2/15	0.60	(1.54)
11. Heptanoic acid	5/15	0.87	(1.45)
12. Limonene	6/15	0.93	(1.44)
13. Hexanoic acid	7/15	0.93	(1.39)
14. Heptane	5/15	1.00	(1.86)
15. Benzyl acetate	7/15	1.40	(2.12)
16. Methyl salicylate	9/15	1.60	(1.86)

Table 7.11 (*Continued*)

Compound	Proportion Detecting	Intensity Mean	Std.
17. β-Ionone	9/15	1.93	(2.21)
18. Anethole	8/15	2.73	(2.86)
19. Heptyl alcohol	13/15	2.80	(1.80)
20. Guaiacol	13/15	2.80	(1.87)
21. Citral	12/15	2.87	(2.25)
22. Camphor	14/15	3.53	(2.09)
23. 4-Methyl valeric acid	9/15	3.93	(3.68)
24. Linalool	13/15	4.00	(2.37)
25. *n*-Butyl ether	13/15	4.00	(2.10)
26. Valeric acid	15/15	5.00	(2.16)
27. 2,4-Pentanedione	15/15	5.57	(1.29)
28. Furfural	15/15	6.07	(1.24)
29. Menthol	15/15	6.14	(0.92)
30. *iso*-Amyl acetate	15/15	6.67	(1.19)
31. *n*-Butyl alcohol	15/15	6.67	(1.30)
32. Acetaldoxime	15/15	6.71	(0.80)
33. 2-Heptanone	15/15	6.73	(1.00)
34. *iso*-Valeric acid	15/15	6.73	(1.24)
35. Ethyl benzene	15/15	6.87	(2.00)
36. *n*-Butyl acetate	15/15	7.33	(1.08)
37. Ethyl acetate	15/15	7.53	(1.02)
38. Methanol	15/15	7.67	(1.14)
39. Benzaldehyde	15/15	7.73	(0.93)
40. Cyclohexanone	15/15	7.80	(1.38)
41. Toluene	15/15	7.87	(1.09)
42. Butyric acid	15/15	7.87	(0.96)
43. Acetal	15/15	8.13	(1.15)
44. Ethyl methyl ketone	15/15	8.40	(0.61)
45. Pyridine	15/15	8.47	(0.72)
46. Acetone	15/15	8.53	(0.88)
47. Propionic acid	15/15	8.73	(0.57)

the compounds had a clear odor to normal observers, and many had been used in previous olfactory research (e.g., references 16 and 34–37). The compounds were of the highest chemical grade commercially available (suppliers: Fisher Scientific and Eastman Kodak) and generally exceeded 99% purity in gas chromatographic analysis done at the Monell Center. A few of the compounds were redistilled to remove contaminates that could have conceivably interfered with the experiment.

Each of the 47 compounds was presented to each observer at neat concentration in 200 ml glass sniff bottles with 5 cm diameter openings. A trial consisted of the random presentation, for approximately 3 seconds, of either a sniff bottle containing the chemical of interest, or a bottle containing an equivalent amount of a control substance, propylene glycol in this case. This was then followed immediately by the presentation of the opposite stimulus (i.e., either odorant or propylene glycol). The observer's task was to state which of the two presentations produced the strongest intranasal sensation. The observer was warned not to sniff the vapor in vigorously, but to first sample it carefully to establish the presence or absence of intranasal sensation. If a clear-cut sensation was not present, the observer was allowed to sniff more vigorously. This procedure ensured that the observer would inhale only small amounts of substances such as pyridine, which produce coughing spasms. It also acted to minimize, at least to some extent, the vapor pressure differences of the full concentration substances entering the nose. Six such detection trials were used for each substance for which an observer did not feel confident that clear-cut intranasal detection had occurred. If an anosmic felt that neither bottle produced an intranasal sensation, he was requested to "guess" one or the other bottle. If an anosmic observer was correct on five of the six trials, it was assumed tuat the stimulus had been detected. Typically, the normals reported detection on the first trial. In any single session, no more than three observers were tested.

Once detection was established, each observer was asked to rate the intensity, pleasantness, coolness, warmth, and presumptive safety on nine point rating scales with the extremes verbally defined as follows: weak–strong; pleasant–unpleasant; cool–warm; and safe–unsafe. The midpoints of the scales were defined as "neither — nor —" for each attribute. To control for possible position biases, approximately half of the observers were tested with scales that had one of its ends (e.g., very pleasant) associated with the right side of the page on which the scale was exhibited, and the others with the scales oriented in the opposite manner (i.e., very pleasant associated with the left side of the page). The responses were tabulated by the experimenter as they were verbalized by the observer. Each odorant was rated by the observer on all the attributes before going on to the next stimulus. The order of presentation of the attribute scales and of the stimulus chemicals was varied randomly from observer to observer, and a minimum of 45 seconds was interspread between the stimulus presentations.

The anosmic group's mean intensity ratings of each compound can be conceptualized as quantitative measures of the interaction between the stimulant and the trigeminal nerve. Data of this nature are ideally suited for regression analysis. Also, the quantitative measurements can be used to divide the data set into discrete groups that could conceivably be separated by discriminant functions. The function could be developed using either multivariate statistical analysis techniques or pattern recognition methods. In this trigeminal study, both approaches were taken using the three different computer analysis methods.

For the regression analysis, a stepwise linear regression program (BMD02R) was utilized (38). This program employs a least squares procedure to fit a linear equation to the dependent variable. For the discriminant function analysis, both a parametric method and a nonparametric method were used. In the parametric study, a stepwise multivariate discriminant analysis program (BMD07M) was employed (38). This program utilized interrelationships among the independent variables to develop a function capable of separating the data into distinct groups. The nonparametric study was done with the linear learning machine, which was described previously.

The molecular descriptors used as independent variables in these computer studies can be divided into two categories: (1) those "readily available" to the average researcher, and (2) those "computer derived." The parameters falling into the first category included physical properties of the compounds available from published tables of experimental data and parameters easily calculated from the compounds' molecular formulae. Molecular weight, vapor pressure, boiling point, the number of oxygen atoms, the number of aromatic ring systems, and Carbowax 20M retention times[1] were all used as the readily available descriptors in these studies. Other parameters (e.g., dipole moment, water solubility) were considered but were excluded because their values were not available from the literature for all the compounds in this work or were not determined under equivalent experimental conditions. While these six parameters used contain information about the molecules' physical properties and chemical nature, the "computer derived" descriptors contain information about the structural nature of the compounds. This second category of parameters included fragment, substructure, environment, molecular connectivity, and geometric descriptors, which are described earlier. Since over 100 descriptors were generated for this study, only those found to be important for discriminating the data set are presented in the following section.

Results for Trigeminal Compounds

The proportion of anosmics detecting the various compounds is presented in Table 7.11 along with the mean intensity rating and its respective standard deviation. The entries in this table are arranged so that the mean intensity

---

[1] Carbowax is the trade name for poly(ethylene glycol), which is used in gas–liquid chromatography as a stationary phase. The 20M designates the average molecular weight of this polymeric material.

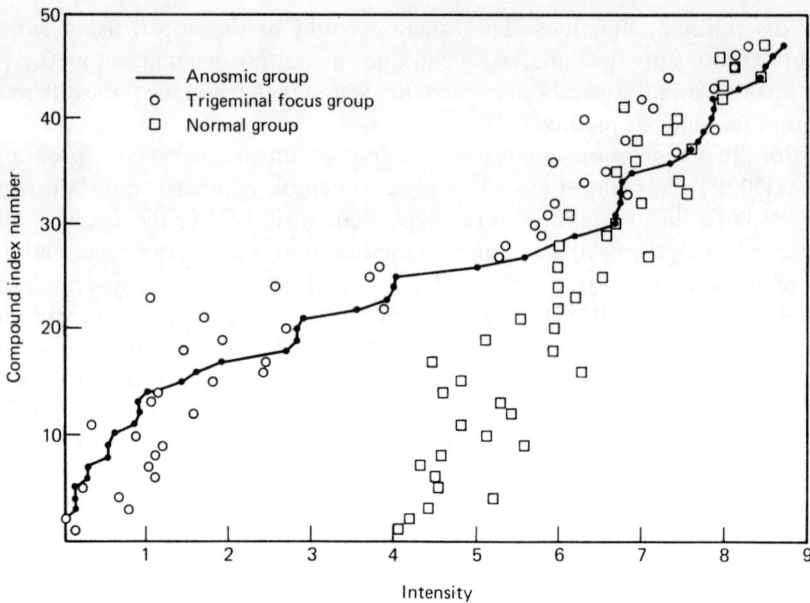

Figure 7.8 A comparison of the average intensity data from the three experimental groups.

ratings increase from top to bottom. The mean intensity scale values presented are based on the ratings from all the observers tested with a zero inserted for observers reporting no detection of a stimulus.

It is apparent from Table 7.11 that a large number of the 47 stimulants were detected by at least some of the individuals lacking olfactory nerve function. Indeed, 45 of the 47 compounds (96%) used in this study were detected by at least one of the anosmics.

In Figure 7.8 the intensity data from the anosmic group are plotted along with each compound's mean intensity rating as determined by each of the other experimental groups. As would be expected, the means of the intensity ratings of both the anosmic and trigeminal focus groups were considerably lower than those of the normal individuals. However, they were strikingly similar for each of the compounds. Another interesting trend is that the intensity ratings for the highly active stimulants are similar for all three groups.

The other psychometric ratings measured in this experiment were found to be clearly intercorrelated with one another, as well as with the intensity ratings. In Table 7.12 the linear correlation coefficients between the four psychometric rating scales are presented. As can be seen, the trends are the same for all three experimental groups. Thus pleasantness and presumptive

Table 7.12 Linear Correlation Coefficients for the Psychometric Rating Scales

	Pleasantness[a]	Coolness	Safety[a]
Anosmics (based on means of 45 stimuli)			
Intensity	−0.84	−0.39[b]	−0.90
Pleasantness	—	0.59[a]	0.91
Cool		—	0.54
Safety			—
Trigeminal Focus (based on means of 47 stimuli)			
Intensity	−0.71	−0.36[c]	−0.70
Pleasantness	—	0.63[a]	0.86
Cool		—	0.54
Safety			—
Normals (based on means of 47 stimuli)			
Intensity	−0.65	−0.46[b]	−0.82
Pleasantness	—	0.81[a]	0.92
Cool		—	0.72
Safety			—

[a] $p < .001$.
[b] $p < .01$.
[c] $p < .05$.

safety are inversely correlated with trigeminal intensity, but directly correlated with one another. The low correlations between the coolness ratings and the other sensations were most likely due to the fact that most of the compounds received near-neutral ratings with only a few of them being rated as markedly cool or warm.

Multiple linear regression analysis was used in an attempt to develop an equation to predict the mean intensity ratings as reported by the anosmic group. First, the five readily available descriptors (molecular weight, vapor pressure, boiling point, the number of oxygen atoms, and the number of aromatic ring systems) were tested. Although all five parameters were included in the equation, the multiple R obtained from the regression of these variables was too small to provide a strong prediction ($R = 0.74$, $s = 2.20$).

In an effort to improve the regression equation, 10 substructure descriptors, reflecting either the branching of the molecule or the presence of functional groups (see Table 7.13), were then regressed. This particular set of substructures was selected because it proved to be useful in preliminary studies

Table 7.13 Substructure Descriptors Used in the Analysis of the Trigeminal Data Set

1.	—OH
2.	—O—
3.	$-\overset{\overset{O}{\|}}{C}-$
4.	$=\overset{\|}{C}=$
5.	$-\overset{\overset{O}{\|}}{C}-O-$
6.	$-\overset{\|}{C}-$
7.	$-\overset{\|}{C}-$
8.	C—C—C—
9.	—C—
10.	—C

of the intensity data using pattern recognition techniques. Although these substructure descriptors were not highly correlated with the intensity data (all correlation coefficients < .40), a regression equation better than the one obtained using the readily available descriptors was found using only five of these substructures ($R = 0.83, s = 1.85$ with substructures 1, 3, 4, 7, and 9 of Table 7.13 as independent variables). The addition of the other five substructures improved the regression equation's multiple R value only negligibly ($R = 0.84, s = 1.85$), possibly by increasing the number of degrees of freedom produced by adding more independent variables to the equation.

When the 5 readily available descriptors and 10 substructure descriptors were combined and regressed, a better equation was found ($R = .89, s = 1.64$) using only 11 variables; the boiling point, the number of aromatic ring systems, and substructures 5 and 6 in Table 7.13 were never included in the equation because they were statistically insignificant. Although twice as many variables were used in this equation, it still did not fit the intensity data as well as desired. Attempts were made to improve the regression results by

Table 7.14 Intensity Data and Physical Constants for the Series of Aliphatic Acids

Acid	Molecular Formula	Calculated log P^a	Boiling Point	Carbowax 20M Retention Time	Average Intensity (Anosmics)
Decanoic	$C_{10}H_{20}O_2$	3.92	269	15.88	0.00
Heptanoic	$C_7H_{14}O_2$	2.33	222	12.79	0.87
Hexanoic	$C_6H_{12}O_2$	1.81	205	11.74	0.93
4-Methyl valeric	$C_6H_{12}O_2$	1.69	200	11.34	3.93
Valeric	$C_5H_{10}O_2$	1.28	185	10.68	5.00
Isovaleric	$C_5H_{10}O_2$	1.16	174	10.09	6.73
Butyric	$C_4H_8O_2$	0.75	162	9.59	7.87
Propionic	$C_3H_6O_2$	0.23	140	8.67	8.73
Correlations with Intensity	Linear	−0.88	−0.91	−0.89	
	Spearman	−1.00	−1.00	−1.00	

^{a} These log P values, which are the partition coefficients of the chemicals between 1-octanol/water, were calculated by a method developed by Rekker et al. (39). A comparison of these calculated values with measured values available in the literature (see reference 40) showed excellent agreement.

using other computer derived and readily available descriptors, but a better regression equation was not found.

The reason for this lack of improvement may be the inadequacy of the molecular parameters tested to encode the large variety of molecular structures and/or the inherent error of the dependent variables as reflected by their large standard deviations. An examination of the homologous series of aliphatic acids contained in the data set (in which the functionality of the molecules is constant) suggested that the variability in the intensity ratings was probably the major factor. In Table 7.14 the aliphatic acids in this data set are arranged by increasing mean intensity as reported by the anosmics. Three physical properties that contain solubility information and the correlations between the perceived intensities and these physical properties are also included. Since the functionality of the acids is constant, the intensities of the compounds should be highly correlated to the solubility properties. Although a perfect Spearman rank–order correlation was obtained in all three cases, the linear correlation coefficients fell into the same absolute range as the multiple R values obtained from the best regression equation. This suggests that the variability (and thus potential error) in the psychometric intensity data probably limits the magnitude of the R values. Therefore, rather than

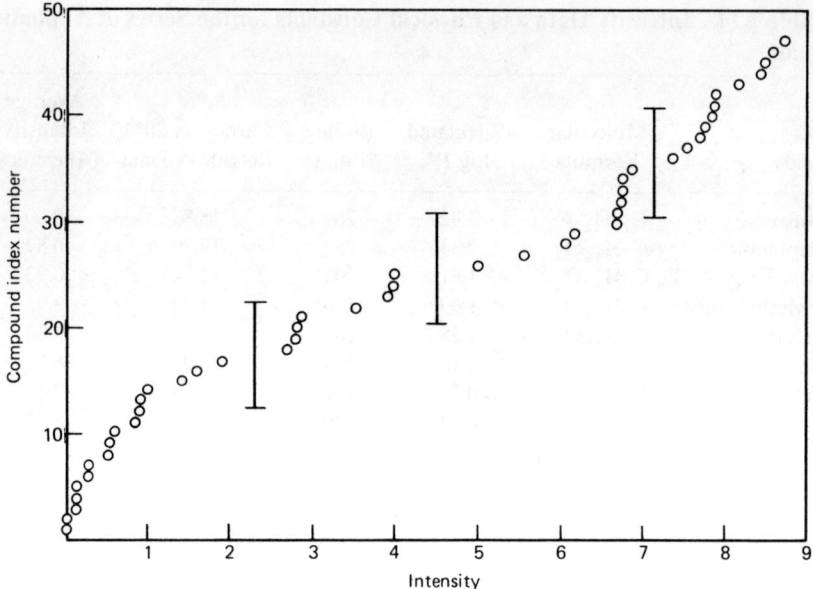

Figure 7.9 Intensity data for anosmics including intensity cutoff points.

continue to search for a better regression equation, discriminant analysis was employed in an attempt to separate the 47 compounds into four intensity classes.

To facilitate the separation of the compounds into groups, a graphical representation of the intensity data for the anosmics was made (see Figure 7.9). The data set was then divided at three points, represented in Figure 7.9 as vertical lines at intensities 2.3, 4.5, and 7.1. These correspond to the low, middle, and high intensity cutoff points, respectively, and were chosen because of the natural gaps in the data at these positions and the limited number of classes into which a data set of 47 members can be validly divided for discriminant analysis.

The development of discriminant functions for each of the intensity cutoff points was treated as three separate two-class problems rather than as a single four-class problem. Thus comparisons among the sets of variables included in each function could be made. The 5 readily available descriptors and the 10 substructure descriptors used in the regression studies were first tested separately and then combined and retested at each intensity cutoff point. From the data given in Table 7.15, it is apparent that, as in the regression studies, the best results were obtained by combining the two different types of descriptors.

Table 7.15 Linear Discriminant Function Analysis Results

Intensity Cutoffs	Descriptor Sets	Number of Descriptors Available	Number of Descriptors Included	Classification Success	
				Below Cutoff	Above Cutoff
Middle	Readily available	5	4	22/3	1/21
	Substructures	10	8	22/3	1/21
	Combined	15	5	24/1	0/22
Low	Readily available	5	5	11/6	5/25
	Substructures	10	6	15/2	3/27
	Combined	15	11	17/0	2/28
High	Readily available	5	1	33/2	5/7
	Substructures	10	9	33/2	3/9
	Combined	15	14	24/1	2/10

At the high intensity cutoff point, the molecular weight alone misclassified the fewest compounds when only the 5 readily available descriptors were tested. The addition of any of the other 4 descriptors did not improve the discrimination. The 10 substructure descriptors produced only slightly better results at this cutoff. However, only 3 compounds (numbers 27, 36, and 42 of Table 7.11) were misclassified when 14 of the 15 variables were incorporated into the discriminant function (substructure 2 was excluded because it was statistically insignificant).

At the low intensity cutoff point, the combined discriminant function classified 45 out of the 47 compounds; only compounds 24 and 25 in Table 7.11 were missed. The 11 independent variables included in this discriminant function were almost identical to the 11 variables used in the combined regression except that substructure 9 was replaced by substructure 5. The reason for this switch is not evident.

Compound 23, 4-methyl valeric acid, was the only compound misclassified at the middle intensity cutoff point. The discriminant function contained five variables: molecular weight and substructures 1, 3, 6, and 8. When the class designation for this compound was switched, the entire data set was classified correctly using the same five variables. The misclassification of 4-methyl valeric acid was not surprising, since its standard deviation was 3.68 intensity units. Thus it is conceivable that it does belong in the high intensity class.

The results of these discriminant analysis studies demonstrated that this data set could be separated into high and low intensity classes. However, total separability was not obtained at each intensity cutoff as expected. Instead of testing a multitude of different variables with discriminant analysis, the nonparametric linear learning machine was used to test the variables used in these two studies, as well as other variables that can be generated in the computer system.

To assess the predictive ability of any set of descriptors developed in this work, 20 randomly selected training and prediction sets were generated from the data set for each cutoff point. Each prediction set contained four data set members randomly drawn from the entire data set. The remaining 43 data set members were used to develop the discriminant function. For the low cutoff point, each prediction set contained one member from below the cutoff point and three members from above the point. This ratio was reversed for the high intensity cutoff point (i.e., three members from below and one member from above). For the middle cutoff point, two members were taken from each side of the cutoff. Each training set was then used to develop a unique discriminant function or weight vector, which was then used to classify the prediction set members omitted during the training. The results from the 20 prediction sets were then averaged to obtain the percent predictive ability.

The readily available and substructure descriptors used previously in the regression and discriminant analysis studies were the first descriptors tested. This was done to provide a comparison of the parametric and nonparametric methods. The readily available descriptors alone were unable to linearly separate the data set at any intensity cutoff point. However, the substructure descriptors were able to separate the data set at the low and middle, but not at the high, cutoff points. The results of these analyses are presented in Table 7.16, where the predictive abilities are average values for the 20 prediction sets described previously.

The poorest results in testing the combined descriptors were obtained at the high intensity cutoff. Although linear separability was obtained using only 10 descriptors (the vapor pressure and substructures 4, 5, 6, and 9 were not included), the predictive ability was only 1.7% above the 75% obtainable by always classifying all prediction set members into the largest intensity class.

At the low intensity cutoff, most of the 15 descriptors were needed to obtain linear separability. The boiling point and substructures 5 and 6 were the descriptors excluded. However, the predictive ability was poor compared to the 75% obtained by guessing the largest class.

The five descriptors found to be the best in the discriminant analysis section (molecular weight, substructures 1, 3, 6, and 8) were unable to separate the data set at the middle cutoff point using the linear learning machine. However, by adding the vapor pressure to the list of descriptors, linear separability

Table 7.16 Results Obtained Using Pattern Recognition Techniques on the Trigeminal Data

Intensity Cutoffs	Descriptor Sets	Number of Descriptors Available	Linearly Separable?	Number of Descriptors Retained	Average Predictive Ability
Low	Readily available	5	No	—	—
	Substructures	10	Yes	9	78.8
	Combined	15	Yes	12	78.8
	A[a]	12	Yes	12	83.8
Middle	Readily available	5	No	—	—
	Substructures	10	Yes	7	86.3
	Combined	15	Yes	6	86.3
	B[a]	13	Yes	13	92.5
High	Readily available	5	No	—	—
	Substructures	10	No	—	—
	Combined	15	Yes	10	76.7
	C[a]	11	Yes	11	86.3

[a] Descriptors identified in Table 7.17.

was obtained. As can be seen, the predictive ability at this cutoff point was well above the 50% value expected from random guessing.

These results clearly indicate that the linear learning machine is better suited than parametric linear discriminant analysis to develop a discriminant function to separate the present data, since linear separability was found in all cases. This is not surprising, since the variables employed in these studies were not independent nor were their covariance matrices equal—assumptions made by the parametric BMD07M program. However, the predictive ability of the various classifiers could conceivably be improved by using differing sets of molecular descriptors. At the low and middle cutoff points, a predictive ability increase between 1 and 2% over the combined descriptors was obtained by adding and substituting computer generated descriptors (i.e., fragment, environment, molecular connectivity, and geometric descriptors) for the readily available descriptors. However, by including the Carbowax 20M retention times, the predictive abilities increased to the values given in Table 7.16 at these two cutoff points. The actual descriptors used in each case are given in Table 7.17. For the high intensity cutoff point, the computer generated descriptors in set C of Table 7.17 increased the predictive ability by 9.6% over the combined descriptors tested previously.

Table 7.17 Descriptors Found to be Most Useful During Pattern Recognition Studies of the Trigeminal Data

Descriptor Set A	Descriptor Set B	Descriptor Set C
1. Number of carbon atoms	1. Number of carbon atoms	1. Number of carbon atoms
2. Longest radius of gyration (X)	2. Longest radius of gyration	2. Number of single bonds
3. Middle radius of gyration (Y)	3. Vapor pressure	3. Number of aromatic bonds
4. Ratio of X radius to Y radius	4. Carbowax 20M retention time	4. Longest radius of gyration
5. Vapor pressure	5. Molecular weight	5. Middle radius of gyration
6. Carbowax 20M retention time	6. —O— environment	6. Molecular weight
7. Molecular weight	7. =C— environment	7. —OH environment
8. =C— environment	8. —OH environment	8. —C— environment
9. —O— environment	9. —C— environment	9. —O— substructure
10. —C= substructure	10. —C— environment	10. =C= substructure
11. =C= substructure	11. —C— substructure	11. C—C—C— substructure
12. C—C—C— substructure	12. =C= substructure	
	13. —C=O substructure	

206

Table 7.18 Descriptors Used for the Prediction of the Olfaction Data Set

Low Cutoff	Middle Cutoff	High Cutoff
1. Number of carbon atoms	1. Number of carbon atoms	1. Number of carbon atoms
2. Longest radius of gyration	2. Longest radius of gyration	2. Longest radius of gyration
3. Number of single bonds	3. Number of single bonds	3. Number of single bonds
4. Number of aromatic bonds	4. Number of aromatic bonds	4. Number of aromatic bonds
5. Molecular weight	5. Molecular weight	5. Molecular weight
6. —C— environment	6. —C— environment	6. Middle radius of gyration
7. —O— environment	7. —O— environment	7. —O— environment
8. —C— environment	8. —C— environment	8. —C— environment
9. —O— substructure	9. —OH environment	9. —OH environment
10. —C= substructure	10. —C— environment	10. —C— environment
11. =C= substructure	11. —C— substructure	11. —O— substructure
12. —C=O substructure	12. —C=O substructure	12. =C= substructure
13. C—C—C— substructure	13. =C= substructure	13. C—C—C— substructure

The descriptor sets in Table 7.17 should not be considered as optimum sets, since they were obtained through heuristic searches. Undoubtedly, there exist several other sets of descriptors that could be found through a systematic combination of the descriptors available, but the computational time involved in such a procedure would likely be impractical. However, the descriptors given in Table 7.17 were the best ones found in this work.

Since classifiers were developed in these studies for separating olfactory stimulants into different trigeminal activity classes, the next logical step was to use them for classifying other odorants. An olfaction data set of 495 compounds was chosen for this study because the compounds were already on computer files, ready to be used, and it was hoped that this study would provide some clue as to which molecular parameters might be important for separating the remaining odor qualities in the olfaction data set.

Although it would have been best to use the descriptors in Table 7.17 for this study, obtaining the vapor pressures and the Carbowax 20M retention times for all 495 data set members proved to be an insurmountable task. Therefore, computer derived descriptor sets, found during the previous studies, were used instead. The sets of descriptors finally selected for each intensity cutoff are shown in Table 7.18. While linear separability was maintained at all three cutoffs, a sacrifice was made in the average predictive ability (low cutoff = 75.0%, middle cutoff = 81.3%, high cutoff = 82.5%). Nevertheless, this study was carried through to the end to see what would happen.

After a weight vector of each intensity cutoff was trained using the entire trigeminal data set, the olfaction data set was classified with these weight vectors into one of the four possible activity categories. Figure 7.10 contains

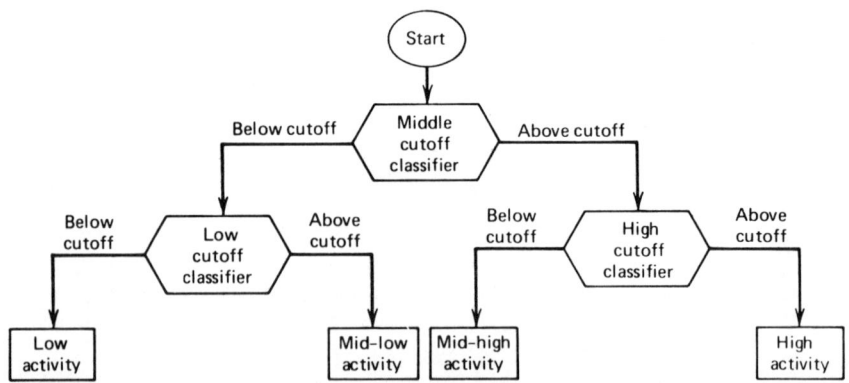

Figure 7.10 The classification scheme used for the prediction of the olfaction data set into trigeminal activity classes.

Table 7.19 Trigeminal Classification Results

Odor Quality	Activity Classes				Percent Conflicts
	Low	Mid–Low	Mid–High	High	
Musk	59	1	0	0	60.9
Floral	34	7	14	9	37.5
Camphor	27	30	13	15	34.5
Minty	17	8	26	10	29.5
Pungent	18	1	12	55	29.1
Putrid	3	1	6	13	17.4
Ethereal	4	1	1	36	33.3
Almond	7	4	7	4	45.4
Aromatic	7	5	1	9	18.2
Aniseed	8	1	0	1	90.0
Lemon	5	0	1	0	0.0
Cedar	5	0	0	0	0.0
Garlic	0	3	1	2	16.7
Rancid	1	0	3	0	0.0

a flow diagram showing the logic used for classifying the odorants. The classification results are reported in Table 7.19. Although a compound needed to go through only two classifiers before being categorized into a group, the third classifier was always consulted to see if its classification was in agreement with the other two (e.g., if both the middle and the high classifiers predicted the compound to be above the cutoff, the low intensity classifier should have also predicted "above"). The number of times a conflict of prediction occurred for each odor quality is shown in the last column of Table 7.19. The number of conflicts occurring in this study was expected, since a small number of compounds was employed for training and the "best" descriptors could not be used.

The seven primary odor qualities in Table 7.19 are arranged according to the predicted trigeminal activity. As can be seen, the more pleasant odor qualities, musk and floral, were primarily predicted as low trigeminal stimulants, while the more unpleasant odorants were classified as high trigeminal stimulants. This was the same relationship found in the study of the 47 trigeminal odorants (see Table 7.12).

As for the classification of each odor quality, only the musk odorants were found to cluster in a single activity class. This was not surprising, since the musk compounds were found to be separable from the other odorants on the basis of a few descriptors. Thus it would appear that the other odor qualities may prove to be difficult to separate.

Discussion of Trigeminal Study

The results of the present study indicate that a large number of volatile chemical compounds, which have odors to normal observers, can be detected intranasally by individuals lacking olfactory nerve function. Furthermore, the results of this study indicate that the perceptual scales of intensity, pleasantness, and safety are systematically interrelated, with the more intense stimuli being rated as more unsafe and unpleasant.

Although the mean intensity ratings of anosmic and trigeminal focus normal observers were, in a general sense, similar (see Figure 7.8), they were not equivalent. The intensity ratings of the trigeminal focus group were larger than the ratings of the anosmics for the less intense stimuli, and smaller than those of the anosmics for the more intense stimuli. While the reason for these differences is not known, one hypothesis worthy of consideration is that the actual physiological response parameters of the trigeminal nerve are equivalent for both the normal and anosmic groups, but that these two groups respond to the trigeminal input differently. Perhaps the task of differentiating trigeminal input from olfactory input is different for the normals for relatively weak trigeminal stimulants, causing some to ascribe aspects of the olfactory stimulation to trigeminal function. Or perhaps, at low concentrations, such individuals are more responsive to demand characteristics of the experimental situation (see reference 41), leading them to report noticing more intense trigeminal stimulation. With the more intense stimuli, the relatively untrained normals may simply be overwhelmed with the overall stimulation, thereby not being able to notice the true nature of the trigeminal input. If this explanation is correct, additional practice or training might produce similar results for both groups.

It is important to note that the present experiment was not designed to establish, on a molecule to molecule basis, the relative ordering of chemicals along a trigeminal stimulation continuum. What it was designed to do was to establish, in a general sense, such an ordering of chemicals when they were presented at maximal concentrations at the external nares in individuals using "natural" sniffing procedures. Thus a sort of mapping of the outer range of the chemicals' trigeminal stimulative properties under normal sniffing conditions was sought. Considerable speculation exists on whether "pure olfactory" stimulants exist (see reference 4), and the use of neat concentrations optimized the possible trigeminal stimulative properties of a given compound, at least before it was drawn into the nares. The results of this study suggest that few pure olfactory stimulants exist when their vapors

Analysis of Trigeminally Active Compounds

are presented to the observer's nostrils at neat concentrations and under natural sniffing conditions. It is conceivable that even "pure air" can produce some trigeminal stimulative effects, depending on its temperature and flow rate. However, the results of the prediction study using the olfaction data set indicated that musk odorants would be the most likely class of compounds to contain "pure olfactory" stimulants.

Although it is likely that vapor pressure is related in some fashion to the production of intranasal trigeminal stimulation, it is unlikely that vapor pressure per se played the major role in producing the intensity differences of the various chemicals used in this study. If this were true, a high correlation would be expected between the perceived intensity and this variable. However, this was not found (linear correlation coefficient = .39). In fact, the boiling point, Carbowax 20M retention times, and molecular weight values all exhibited higher linear correlations to the intensity data than did the vapor pressure ($-.60$, $-.58$, and $-.70$, respectively). It is probable that the variations in sniffing and the intake of chemicals differentially influenced the concentrations of the various stimuli reaching the epithelial regions. Indeed, it is likely that the reflexive or conscious sniffing strategies utilized by the observers minimized the intake of highly volatile vapors and maximized the intake of less volatile ones, acting somewhat analogously to the pupil of the eye in controlling stimulus input.

While the regression procedure was not particularly successful in predicting the perceived intensity, the results of the discriminant analysis and the pattern recognition studies clearly indicated that this data set could be divided into four intensity groups using only a few molecular parameters. Of the two methods, the nonparametric pattern recognition method produced the best results, with discriminant functions being developed at all three intensity cutoff points to separate the data completely.

Of the many different molecular parameters tested in this experiment, no single one could account for the intensity measurements of all 47 data set members. Furthermore, the molecular parameters found to give the highest performance at each intensity cutoff were not identical. Molecular weight, the number of carbon atoms, the longest radius of gyration, and substructure 4 of Table 7.13 were the only descriptors included in those given in Table 7.17. The rest of the descriptors were similar in that they contained information about the molecule's structural composition and its physical properties. These results and the small data set make it impossible to list an exact set of descriptors necessary to predict the trigeminal intensity. Despite this fact, however, an important trend was present. In every computer study done in this work, the ability of the classifier to distinguish between the high and low intensity compounds at each of the cutoff points improved when information about the type of functional groups present in the compound was added to the readily available descriptors containing primarily information about the molecules' physical properties. This finding supports the hypothesis that trigeminal nerve stimulation involves two steps: (a) transport of the molecule from the air to the nerve ending through the mucus layer, and (b) the molecular interaction at the nerve ending itself. For the first step, water solubility, lipid solubility, vapor pressure, and Carbowax 20M retention times would all give an indication of the ability of the molecule to reach the nerve endings. These relationships were suggested in the aliphatic acid series given in Table 7.14, where strong correlations were present between perceived intensity and these physical properties. However, since several different molecules can have the same ability to reach the trigeminal nerve (i.e., have nearly identical solubility parameters), the differences in the perceived intensities are presumably due to other molecular properties. These data suggest that the functional groups present in the molecule may be an important factor, although additional data from a larger set of compounds are needed before this notion can be adequately verified. In the future, the specific factors involved within each of the processes will hopefully be delineated.

This study and the previous one on musk odorants have clearly demonstrated the utility of pattern recognition techniques in studies of structure–activity relations. In both cases a large number of descriptors were analyzed with a meaningful subset being identified at the conclusion of each study. These studies have uncovered new structural relationships. This information will provide answers to some questions leading no doubt, to new hypotheses.

REFERENCES

1. W. H. Hollinshead, *Textbook of Anatomy*, Harper and Row, New York, 1974.
2. D. G. Moulton, The Olfactory Pigment, in *Handbook of Sensory Physiology*, Vol. IV, *Chemical Senses*, Part 1, *Olfaction*, L. M. Beidler (Ed.), Springer-Verlag, Berlin, 1971.
3. P. P. C. Graziadei, The Olfactory Mucosa of Vertebrates, in *Handbook of Sensory Physiology*, Vol. IV, *Chemical Senses*, Part 1, *Olfaction*, L. M. Beidelr (Ed.), Springer-Verlag, Berlin, 1971.
4. D. Tucker, Nonolfactory Responses from the Nasal Cavity, Jacobson's Organ and the Trigeminal System, in *Handbook of Sensory Physiology*, Vol. IV, *Chemical Senses*, Part 1, *Olfaction*, L. M. Beidler (Ed.), Springer-Verlag, Berlin, 1971.
5. D. Ottoson, The Electro-Olfactogram, in *Handbook of Sensory Physiology*, Vol. IV, *Chemical Senses*, Part 1, *Olfaction*, L. M. Beidler (Ed.), Springer-Verlag, Berlin, 1971.
6. R. C. Gesteland, Neural Coding in Olfactory Receptor Cells, in *Handbook of Sensory Physiology*, Vol. IV, *Chemical Senses*, Part 1, *Olfaction*, L. M. Beidler (Ed.), Springer-Verlag, Berlin, 1971.
7. T. C. Lucretius, *The Nature of the Universe*, 47 BC, transl. by Latham, Penguin Books, London, 1951.

References

8. R. W. Moncrieff, *The Chemical Senses*, 2nd ed., Leonard Hill Ltd., London, 1951.
9. G. M. Dyson, Raman Effect and the Concept of Odour, *Perfumery Essent. Oil Rec.*, **28**, 13 (1937).
10. R. H. Wright, Odour and Molecular Vibration. I. Quantum and Thermodynamic Considerations, *J. Appl. Chem.*, **4**, 611 (1954).
11. R. H. Wright and R. S. E. Serenius, Odour and Molecular Vibration. II. Raman Spectra of Substances with the Nitrobenzene Odour, *J. Appl. Chem.*, **4**, 615 (1954).
12. R. H. Wright and R. E. Burgess, Musk Odour and Far Infrared Vibration, *Nature (Lond.)*, **224**, 1033 (1969).
13. J. E. Amoore, The Stereochemical Specificities of Human Olfactory Receptors, *Perfumery Essent. Oil Rec.*, **43**, 321 (1952).
14. J. E. Amoore, J. W. Johnston, Jr., and Martin Rubin, The Stereochemical Theory of Odor, *Sci. Am.*, **210**, 42 (1964).
15. M. G. J. Beets, in *Molecular Structure and Organoleptic Quality*, S.C.I. Monograph No. 1, Society of Chemistry and Industry, London, 1957.
16. S S. Schiffman, Physicochemical Correlates of Olfactory Quality, *Science*, **185**, 112 (1974).
17. H. Boelens, Molecular Structure and Olfactive Properties, in *Structure–Activity Relationships in Chemoreception*, G. Benz (Ed.), Information Retrieval Ltd., London, 1976.
18. E. T. Theimer and J. T. Davies, Olfaction, Musk Odor, and Molecular Properties, *J. Agr. Food Chem.*, **15**, 6 (1967).
19. A. Dravniek and P. Laffort, Physico-Chemical Basis of Quantitative and Qualitative Odor Discrimination in Humans, in *Olfaction and Taste IV*, D. Schneider (Ed.), Wissens-Verlag-MBH, Stuttgart, Germany, 1972.
20. T. F. Wood, *The Givandan*, nine papers, January 1968 to April 1970.
21. J. E. Amoore, *Molecular Basis of Odor*, Charles C. Thomas, Springfield, Ill., 1970.
22. T. J. Kulle and P. G. Cooper, Effects of Formaldehyde and Ozone on the Trigeminal Nasal Sensory System, *Arch. Environ. Med.*, **30**, 237 (1975).
23. S. D. Murphy, H. V. Davis, and V. L. Zaratzian, Biochemical Effects in Rats from Irritating Air Contaminants, *Toxicol. Appl. Pharmacol.*, **6**, 520 (1964).
24. H. Salem and H. Cullumbine, Inhalation Toxicities of Some Aldehydes, *Toxicol. Appl. Pharmacol.*, **2**, 183 (1960).
25. W. F. Allen, Effect on Respiration, Blood Pressure, and Carotid Pulse of Various Inhaled and Insufflated Vapors When Stimulating One Cranial Nerve and Various Combinations of Cranial Nerves, *Am. J. Physiol.*, **87**, 319 (1928).
26. W. F. Allen, Effect of Various Inhaled Vapors on Respiration and Blood Pressure in Anesthetized, Unanesthetized, Sleeping, and Anosmic Subjects, *Am. J. Physiol.*, **88**, 620 (1929).
27. W. F. Allen, Olfactory and Trigeminal Conditioned Reflexes in Dogs, *Am. J. Physiol.*, **118**, 532 (1937).
28. W. S. Cain, Contribution of the Trigeminal Nerve to Perceived Odor Magnitude, *Ann. N.Y. Acad. Sci.*, **237**, 28 (1974).
29. R. L. Doty, Intranasal Trigeminal Detection of Chemical Vapors by Humans, *Physiol. Behav.*, **14**, 855 (1975).
30. W. S. Cain, Olfaction and the Common Chemical Sense: Some Psychophysical Contrasts, *Sensory Processes*, **1**, 57 (1976).

31. J. F. Kallman, W. A. Schonfeld, and S. E. Barrera, Genetic Aspects of Primary Eunuchoidism, *Am. J. Ment. Defic.*, **48**, 203 (1944).
32. S. S. Stevens and E. H. Galanter, Ratio and Category Scales for a Dozen Perceptual Continua, *J. Exp. Psychol.*, **54**, 337 (1957).
33. N. I. Sax, *Dangerous Properties of Industrial Materials*, Reynold, New York, 1966.
34. B. U. Berglund, U. Berglund, G. Ekman, and T. Engen, Individual Psychophysical Functions for 28 Odorants, *Percept. Psychophys.*, **9**, 379 (1971).
35. R. L. Doty, An Examination of Relationships between the Pleasantness, Intensity, and Concentration of 10 Odorous Stimuli, *Percept. Psychophys.*, **17**, 492 (1975).
36. T. Engen and C. O. Lindstrom, Psychophysical Scales of the Odor Intensity of Amyl Acetate, *Scand. J. Psychol.*, **4**, 23 (1963).
37. H. R. Moskowitz, A. Dravneiks, and C. Gerbers, Odor Intensity and Pleasantness of Butanol, *J. Exp. Psychol.*, **103**, 216 (1974).
38. W. J. Dixon (ed.), *BMD Biomedical Computer Programs*, University of California Press, Los Angeles, Calif., 1973.
39. G. G. Nys and R. F. Rekker, Statistical Analysis of a Series of Partition Coefficients with Special Reference to the Predictability of Folding of Drug Molecules. Introduction of Hydrophobic Fragmental Constants (f values), *Chim. Ther.*, **8**, 521 (1973).
40. A. Leo, C. Hansch, and D. Elkins, Partition Coefficients and Their Uses, *Chem. Rev.*, **71**, 525 (1971).
41. M. T. Orne, On the Social Psychology of the Psychological Experiment: With Particular Reference to Demand Characteristics and Their Implications, *Am. Psychol.*, **17**, 776 (1962).

APPENDIX

A List of the Musk Odorants

1. Allopregnan-3α-ol
2. Androstan-3α-ol
3. Androstan-3β-ol
4. Δ^{16}-Androsten-3α-ol
5. 2-Bromo-4-t-butyl-5-methoxytoluene
6. 2-Bromo-4,6-dinitro-1,3-dimethyl-5-t-butylbenzene
7. Cycloheptadecanone
8. Δ^9-Cycloheptadecen-1-one (Civettone)
9. Cyclohexadecanone
10. Cyclooctadecanone
11. Cyclopentadecanone (Exaltone)
12. Cyclotetradecanone
13. Decamethylene malonate
14. Decamethylene oxalate
15. 4,6-Dinitro-2-azido-1,3-dimethyl-5-t-butylbenzene
16. 3,5-Dinitro-2,4-dimethyl-6-t-butylacetophenone
17. 3,5-Dinitro-2,4-dimethyl-6-t-butylbenzaldehyde
18. 3,5-Dinitro-2,6-dimethyl-4-t-butylbenzonitrile
19. 2,4-Dinitro-3,5-dimethyl-6-fluoro-t-butylbenzene
20. 2,6-Dinitro-3,5-dimethyl-4-fluoro-t-butylbenzene
21. 3,5-Dinitro-2-methyl-4-methoxyacetophenone
22. 4,6-Dinitro-2,3,5-trimethyl-t-butylbenzene
23. Dodecamethylene carbonate
24. Dodecanedicarboxylic acid anhydride
25. α-Dodecyl-γ-butyrolactone
26. Ethylene undecanedioate
27. Δ^{16}-Etiocholen-3β-ol
28. α-Geranyl-γ-butyrolactone
29. α-Heptyl-γ-butyrolactone
30. Hexadecamethylene imine
31. Hexadecanedicarboxylic acid anhydride

32. Hexadecanolactone
33. Δ^7-Hexadecanolactone (Ambrettolide)
34. D-Homoandrostan-3α-ol
35. 3-Methylandrostan-3α-ol
36. 3-Methylandrostan-3β-ol
37. 17-Methylandrostan-3α-ol
38. 17-Methylandrostan-3β-ol
39. 3-Methylcyclopentadecanone (Muscone)
40. 1-Methylcyclopentadecan-2-one
41. 1-Methylcyclopentadecan-4-one
42. A-Norandrostan-2α-ol
43. γ-Octyl-γ-butyrolactone
44. Pentadecanolactone
45. Phenylacetic acid
46. α-Rhodinyl-γ-butyrolactone
47. Tetradecamethylene carbonate
48. Tetradecanolactone
49. Tridecamethylene carbonate
50. Tridecanolactone
51. 2,4,6-Trinitro-3,5-dimethyl-*t*-butylbenzene
52. 2,4,6-Trinitro-3-methyl-5-bromo-*t*-butylbenzene
53. 2,4,6-Trinitro-3-methyl-*t*-butylbenzene
54. 2,4,6-Trinitro-3-methyl-5-chloro-*t*-butylbenzene
55. 2,4,6-Trinitro-3-methyl-1,5-di-*t*-butylbenzene
56. 2,4,6-Trinitro-3-methyl-5-fluoro-*t*-butylbenzene
57. 2,4,6-Trinitro-1-methyl-3-*n*-hexylbenzene
58. 2,4,6-Trinitro-3-methyl-5-iodo-*t*-butylbenzene
59. 2,4,6-Trinitro-3-methylisopropylbenzene
60. Undecamethylene oxalate

Index

Ab initio calculations, 17
ADAPT, 126
Adaptive pattern search, 87
Additivity model, 15
Anosmia, 193
Anosmics, 193-194
Atom-by-atom searching, 76
Atom descriptors, 74
Atom fragment descriptors, 74
Autocorrelation matrix, 44
Autoscaling, 38, 145, 157

Barbiturates, 151
Bayes classifier, 52-54
Bayes formula, 52
Bayesian statistics, 52
Bond descriptors, 75
Biological effectiveness, 2
Biological reaction rate, 3-4, 15
Bond fragment descriptor, 74
Branching index, 81-82

Canonical connection table, 68
Chemical senses, 170
Chemical structure information handling, 62
Class conditional probability, 52
Classification, 31, 51
  definition of, 31, 52
CLSMKR, 131
Clustering, 39, 56
Clustering algorithms, types of, 56-57
Clustering transformations, 39-41
CNDO, 17
COLATE, 135
Complete neglect of differential overlap (CNDO), 17
Conditional probability, 52-53

Confidence limit for regression parameters, 13
Conformations, 18
Congeneric series, 3
Connection tables, 63
  properties of, 66-68
  uniqueness, 68
CORCOF, 135
Correlation coefficient, 47
Covariance, 43, 46, 49
Criterion functions, 50, 100-101

DAN number, 130
Deadzone, 99-100, 148
Decision function, 53, 54
Decision surface, 98
Descriptor development, 62, 131
DEXTR, 134
Dichotomization ability, 106
Discriminant function, 53
  analysis, 197, 202-205
  development of, 95
Discrimination ability, 164
Divergence, 46
DMCON, 133
DMFRAG, 132
DMGEO, 134
DMSS, 133
DMVOL, 134
Drug receptor, 18
Drug structure activity studies, 138

E_s, 3, 8
ED_{50}, 3
Electronic parameters, 7
Electronic properties, 3
Entropy minimization, 45-46

Environment descriptors, 73, 79-81, 92, 156, 197
Error correction feedback, 96-99
Expectation value, 121
Explained variance, 12
Extended Hückel method, 17
External descriptors, 134, 135

F statistic, 9, 12, 46
F-test in regression analysis, 12
Feature extractor, 33
Feature selection, 37, 41-48, 112
 definition of, 37, 112
 types of, 41, 42
Fifth cranial nerve, 172
First cranial nerve, 172
Fisher ratio, 46
Forward selection, 13
Fragment descriptors, 73-75, 92, 132-133, 143, 156, 177, 197
Free-Wilson analysis, procedure for, 16-17
Free-Wilson model, 15

Geometric descriptors, 73, 90-92, 93, 178, 197

Hammett constants, 3, 7-8
Hammett equation, 3
Hammett parameters, 8
Hansch analysis, 2
 basic assumptions of, 3
 linear model for, 3
 quadratic model for, 7
 for studying olfactory stimuli, 175
Highest occupied molecular orbital (HOMO), 17
Ho-Kashyap algorithm, 102-103
Hückel method, 17
Hydrophobic character, 7
Hydrophobic fragment constants, 5
Hydrophobic parameters, 5
Hydrophobic properties, 3

I_{50}, 3
Indicator variables, 9
Intrinsic variables, 109
ISODATA clustering algorithm, 57

Karhunen-Loeve transform, 44, 48
Kinetics, transport, 6

LD_{50}, 3
Lead compound, 1
 discovery of, 1
 exploration of, 2
Least mean square error procedure, 102, 103
Leave-one-out procedure, 146, 162
Linear decision functions, 95
Linear discriminant functions, 96
Linear free energy approach, 2-3. *See also* Hansch analysis
Linear learning machine, 96-100
 limitations of, 100
Linear model, 3
Linear notation, 63, 73
 properties of, 63
 specificity of, 63
 uniqueness of, 63
Lipophilicity, calculation of, 5
Log P, 5
Lowest empty molecular orbital, 17

MATH, 135
Measurement space, 36
MIC, 3
MINDO, 17
Molar refractivity, 9
Molecular conformation, low energy, 18, 83-90
Molecular connectivity, 81, 133
Molecular connectivity descriptor, 73, 81-82, 93, 133-134, 157, 197
Molecular descriptors, 62
Molecular fragments, 132
Molecular geometry, 134
Molecular mechanics, 83-90
Molecular modelling, 83
Molecular orbital theory, 17
Molecular volume, 9, 91-92, 134
Molecular weight, 9, 197
MOLMEC, 83
Multidimensional scaling, 49-51
Multiple correlation coefficient, 11
Multisource measurements, 36
Musk ordorants, 176

Index

N-dimensional space, 36
Nearest neighbor rule, 104-105
Negative feedback, 96
N-octanol, 3
Non-linear mapping, 49-51
   criterion functions for, 50
Nonparametric classifier, 97
Nonparametric methods, 55, 95
Normal distribution, 10
Normalization, 38

Odor quality, 176
Olfactory cells, 172
Olfactory epithelium, 171
Olfactory nerve, 172
Olfactory stimulants, 170

Parallel tangents, 87
Parametric classification, 52
Parametric methods, 55, 95
Partial regression coefficients, 10
Partition coefficient, 3, 5, 201
Partitioning model, 7
Pattern recognition applications, 19
Pattern recognition techniques, attributes of, 31
   basic steps for applying, 35
PCILO, 17
Pharmacophores, 23
Pi (partition coefficient), 3, 5, 201
Prediction set, 99
Predictive ability, of discriminant function, 109-112, 164
   measures of, 47
Preprocessing, 38
Primary odors, 174
Principle moments, 90
Problems in data handling, 128
Psychometric ratings, 198
Psychotropic agents, 139

Quadratic model, 6
Quantum mechanical methods, 17-18

Radius of gyration, 90
Random classifications, 107-112
Recognition, definition of, 31, 37
Regression analysis, 3, 9

applications of, 14, 175, 199-201
dependent variable in, 9
independent variable in, 9
limitations of, 14
types of algorithms for, 13
Regression coefficient, 9
Relative variation, 120, 122, 124
Ridge regression, 13
R_m value, 7

Scaling, 38
Sedatives, 139
Set reduction, 76
SFILES, 129
Sigma, Hammett's, 3, 8
Similarity, 30
Single feature statistics, 47
Single source measurements, 36
Standard deviation of regression parameters, 12
Standard error of regression, 11
Statistical parameters, 9
Steepest descent, 87
Steric parameters, 8
Steric properties, 3
Strain energy function, 84
Strain function, terms in 84-87
Structure encoding, 68
Student t-test, 13
Substructure descriptors, 73, 75-78, 92, 133, 143, 156, 177
Substructure searching, 76, 133
Substructure unit, common musk, 186-192

Taft steric constant, 3, 8-9
Three dimensional models, 22, 83-90
Topological descriptors, 73
Topological representations, 63
Training procedure, 98
Training set, 37
Tranquilizers, 139
Transducer, chemical, 36
   definition of, 36
Trigeminal stimulants, 192

UDRAW, 69, 129
U-statistic, 46

Van der Waals radius, 9, 91
Van der Waals volume, 91
Variance, 9, 121
Variance feature selection, 113-125, 149, 160
Variance weighting, 39, 145
Vomeronasal epithelium, 172

Weight-sign feature selection, 114-115, 148-149
Weight vector, 98
Whole molecule effect, 173
Wiswesser line notation, 64-66